Offline First Web Development

Design and implement a robust offline app experience
using Sencha Touch and PouchDB

Daniel Sauble

PUBLISHING

BIRMINGHAM - MUMBAI

Offline First Web Development

First published: November 2015

Production reference: 1131115

Published by Packt Publishing Ltd.
Livery Place
35 Livery Street
Birmingham B3 2PB, UK.

ISBN 978-1-78588-457-3

www.packtpub.com

Credits

Author
Daniel Sauble

Reviewers
Abu S. Kamruzzaman
Ahmad Nassri

Commissioning Editor
Neil Alexander

Acquisition Editor
Manish Nainani

Content Development Editor
Ritika Singh

Technical Editor
Suwarna Patil

Copy Editor
Tasneem Fatehi

Project Coordinator
Judie Jose

Proofreader
Safis Editing

Indexer
Mariammal Chettiyar

Graphics
Disha Haria

Production Coordinator
Arvindkumar Gupta

Cover Work
Arvindkumar Gupta

About the Author

Daniel Sauble is part UX designer, part developer, and part researcher. He loves enterprise software start-ups and has worked at companies including Puppet Labs and Sonatype on problems encompassing configuration management, repository management, and patch management. Ironically, his first foray into book authorship has nothing to do with any of these.

In his leisure time, he runs, speaks, writes, and spends time with his family. He has learned that there is nothing more painful than the end of an ultramarathon, more nerve-racking than your first conference talk, or more satisfying than a long writing project. One day, he may be foolish enough to start his own company, but for now, he is content to hone his product designing skills in the midst of the start-up culture.

Home is the verdant landscape of the Pacific Northwest, but Daniel enjoys a bit of travel now and then. Between travel, family, work projects, and the personality of an INTJ, he doesn't have much of a social life. He has no illusions that writing a book will change this much.

I would like to thank my wife, Audrey, for bearing with my weekends and evenings spent on this project. I would also like to thank my son for his patience and forbearance while his Papa spent so much time writing this thing called a "book". Not least, I would like to thank Randall Hansen for introducing me to UX design and pushing me to write, speak, and be a better person.

About the Reviewers

Abu S. Kamruzzaman has been a web programmer and database analyst since 2004 and teaching various IT courses as an adjunct lecturer since 2001. He has an interest in a wide range of technologies, such as Hadoop, Hive, Pig, NoSQL databases, Java, cloud computing, mobile app development, and so on. He has vast experience in application development and database implementation for the education sector. He currently works full-time for CUNY (the City University of New York) Central as a PeopleSoft development specialist since November, 2014. His current project is working with the business intelligence team to build a data warehouse for CUNY using OBIEE. Before joining CUNY Central, he worked in various CUNY campuses since 2001. Aside from his full-time job, he has been teaching graduate and undergraduate courses on J2EE, DBMS, data warehousing, object-oriented programming, web design, and web programming in different CUNY campuses since 2001. He is currently teaching for the CIS department under the Zicklin School of Business at Baruch College/CUNY. He enjoys learning new technologies and solving complex computing problems and spends his leisure time doing community work. He has a great interest in open source technologies and teaching students through his lectures. He is also interested in writing books soon but is waiting on choosing the suitable topics of his best interest. He has a master's degree from Brooklyn College (CUNY) and achieved a bachelor's degree in computer science from Binghamton University (SUNY). His web address is `http://faculty.baruch.cuny.edu/akamruzzaman/`.

Abu has worked on the book, *Mastering Spring MVC 4*, by Packt Publishing.

> I wanted to thank my sweet and beautiful wife, Nicole Woods,
> Esq. for her constant patience, support, and encouragement in
> everything I do. I would also like to thank the author and the
> publishing team for giving me an opportunity to be part of
> the team working on this book.

Ahmad Nassri is an advocate of all things open source and a developer tooling enthusiast.

Currently, he is leading the talented engineering team at Mashape, powering API-driven software development for engineering and IT teams across the world. In his spare time, Ahmad blogs on technology and leadership, mentors early stage start-ups, and builds open source projects used by thousands of developers worldwide.

He also speaks on open source, leadership, and the hacker culture in technology conferences. You can check out his speaking schedule and read his latest posts on his blog at `https://www.ahmadnassri.com/`.

I wanted to thank my amazing wife, Nicole, for her continuous love, support, and–most importantly–her patience and encouragement while I burn the midnight oil working on open source project initiatives. I would also like to thank the author and the publishing team for the opportunity to collaborate on this book.

www.PacktPub.com

Support files, eBooks, discount offers, and more

For support files and downloads related to your book, please visit www.PacktPub.com.

Did you know that Packt offers eBook versions of every book published, with PDF and ePub files available? You can upgrade to the eBook version at www.PacktPub.com and as a print book customer, you are entitled to a discount on the eBook copy. Get in touch with us at service@packtpub.com for more details.

At www.PacktPub.com, you can also read a collection of free technical articles, sign up for a range of free newsletters and receive exclusive discounts and offers on Packt books and eBooks.

https://www2.packtpub.com/books/subscription/packtlib

Do you need instant solutions to your IT questions? PacktLib is Packt's online digital book library. Here, you can search, access, and read Packt's entire library of books.

Why subscribe?

- Fully searchable across every book published by Packt
- Copy and paste, print, and bookmark content
- On demand and accessible via a web browser

Free access for Packt account holders

If you have an account with Packt at www.PacktPub.com, you can use this to access PacktLib today and view 9 entirely free books. Simply use your login credentials for immediate access.

Table of Contents

Preface

Do you want to build resilient online apps that delight and satisfy users, regardless of the network state? Do you want to learn about frameworks that make cross-device synchronization easy? Do you need to communicate the value and efficacy of designing for everyone and not just people with a fast, reliable Internet connection?

If so, this book is for you. We'll start by talking about the realities of poor and non-existent Internet connections, then develop an offline app using Sencha Touch, and gradually convert it to a fully functional online experience that retains the robustness of its offline roots. You'll learn valuable tips and techniques that you can use in your own projects.

What this book covers

Chapter 1, *The Pain of Being Offline*, illustrates the pain of a typical offline experience with a story. It explains how this is caused by not designing for the worst-case scenario; in this case, a lack of Internet connectivity. It proposes that a better way of developing web apps is to assume that a reliable Internet connection will not always be available and hence, build to this assumption.

Chapter 2, *Building a To-do App*, takes you through the setting up of your development environment. It implements a basic offline version of the to-do app using Sencha Touch. It describes the principles of a great offline experience and summarizes the ways in which the app must evolve in an online context to continue to adhere to those principles.

Chapter 3, *Designing Online Behavior*, talks about using offline behavior as a guide and planning the online experience of the app. What superset of features will the to-do app support when online? Think about the app in terms of a bunker versus beach house analogy.

Chapter 4, Getting Online, discusses the available offline databases (including PouchDB, remoteStorage, and Hoodie) and switches the app storage to PouchDB. It shows you how to use IBM Cloudant to host and distribute the data in your app.

Chapter 5, Be Honest about What's Happening, informs the user whether the to-do app is offline or online. It assures the users that the changes that they make while offline will be eventually saved, so they don't need to worry about data loss. When the network is flaky, adjust the behavior to compensate, inform the user of the trouble, and let them provide direction.

Chapter 6, Be Eventually Consistent, takes you through the split-brain problem that occurs in networks when two databases lose connectivity, resulting in different data in each database. Sometimes, these conflicts can be resolved automatically when the connection is restored, but most often, it's not. Make the to-do app handle these situations gracefully.

Chapter 7, Choosing Intelligent Defaults, investigates where our defaults fall short and how to improve them. Up to now, we've stuck with the standard defaults provided by our offline database. Learn to recognize the scenarios where our defaults are insufficient and ask (or infer from) the users to guide the caching behavior. Improve the empty states and error messages in our app.

Chapter 8, Networking While Offline, talks about the ways we can extend the illusion of being online through these additional mechanisms: mesh networking, peer-to-peer sharing, and spatial context. When offline, it's possible for a device to network with other devices via Wi-Fi or Bluetooth.

Chapter 9, Testing and Measuring the UX, takes you through creating a test for the to-do app, showing how it performs under different network conditions. It runs both the offline app and online app, comparing the behavior of each. It shows you how the new app retains the robustness of the offline version, while adding the flexibility of online functionality. It suggests the ways in which it could be further improved and ends with suggestions for further reading.

What you need for this book

A laptop or desktop and (optionally) a mobile device running one of the major mobile operating systems. The instructions in this book assume that you use the Chrome browser and Apple products, but the development tools are cross-platform, so any of the major browsers or operating systems should suffice.

If you want to deploy your app to an iOS device, you will need to purchase an Apple Developer membership. We will discuss the details of this in *Chapter 2, Building a To-do App*.

Who this book is for

This book is for mobile application designers and developers who work with applications that rely heavily on Internet connectivity and who would like to improve the robustness and ease of use of their applications through the adoption of the offline-first paradigm.

Conventions

In this book, you will find a number of styles of text that distinguish between different kinds of information. Here are some examples of these styles, and an explanation of their meaning.

Code words in text, database table names, folder names, filenames, file extensions, pathnames, URLs, and user input are shown as follows: Copy this file (named pouchdb-3.6.0.min.js or similar) to the todo-app/ folder.

A block of code is set as follows:

```
Ext.define('TodoApp.store.Item', {
  extend: 'Ext.data.Store',
  config: {
    model: 'TodoApp.model.Item',
    autoLoad: true
  }
});
```

Any command-line input or output is written as follows:

```
$ brew install node
```

New terms and **important words** are shown in bold. Words that you see on the screen, for example, in menus or dialog boxes, appear in the text like this: "Clicking the **Next** button moves you to the next screen."

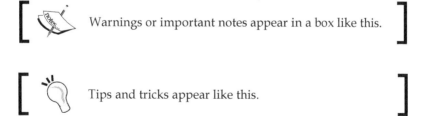

Warnings or important notes appear in a box like this.

Tips and tricks appear like this.

Reader feedback

Feedback from our readers is always welcome. Let us know what you think about this book—what you liked or disliked. Reader feedback is important for us as it helps us develop titles that you will really get the most out of.

To send us general feedback, simply e-mail feedback@packtpub.com, and mention the book's title in the subject of your message.

If there is a topic that you have expertise in and you are interested in either writing or contributing to a book, see our author guide at www.packtpub.com/authors.

Customer support

Now that you are the proud owner of a Packt book, we have a number of things to help you to get the most from your purchase.

Downloading the example code

You can download the example code files from your account at http://www.packtpub.com for all the Packt Publishing books you have purchased. If you purchased this book elsewhere, you can visit http://www.packtpub.com/support and register to have the files e-mailed directly to you.

Downloading the color images of this book

We also provide you with a PDF file that has color images of the screenshots/diagrams used in this book. The color images will help you better understand the changes in the output. You can download this file from https://www.packtpub.com/sites/default/files/downloads/4573OT_ColouredImages.pdf.

Errata

Although we have taken every care to ensure the accuracy of our content, mistakes do happen. If you find a mistake in one of our books—maybe a mistake in the text or the code—we would be grateful if you could report this to us. By doing so, you can save other readers from frustration and help us improve subsequent versions of this book. If you find any errata, please report them by visiting http://www.packtpub.com/submit-errata, selecting your book, clicking on the **Errata Submission Form** link, and entering the details of your errata. Once your errata are verified, your submission will be accepted and the errata will be uploaded to our website or added to any list of existing errata under the Errata section of that title.

To view the previously submitted errata, go to `https://www.packtpub.com/books/content/support` and enter the name of the book in the search field. The required information will appear under the **Errata** section.

Piracy

Piracy of copyrighted material on the Internet is an ongoing problem across all media. At Packt, we take the protection of our copyright and licenses very seriously. If you come across any illegal copies of our works in any form on the Internet, please provide us with the location address or website name immediately so that we can pursue a remedy.

Please contact us at `copyright@packtpub.com` with a link to the suspected pirated material.

We appreciate your help in protecting our authors and our ability to bring you valuable content.

Questions

If you have a problem with any aspect of this book, you can contact us at `questions@packtpub.com`, and we will do our best to address the problem.

1
The Pain of Being Offline

Susan Hopkins is the VP of design at a tech start-up in Silicon Valley. At 63 employees and counting, her day is as much about hiring and culture as it is about design. In her 15-year career, she's seen plenty of both.

Right now, she's shepherding the design of the company's first 1.0 product. There are a lot of moving parts shared among product, design, development, marketing, sales, documentation, and support. She has to wear a lot of hats. She loves every minute of it, but it's a challenging job.

Fortunately, Susan has to-do lists. Everything in her life is organized in this way: shopping, work, personal projects, gift ideas, movies to watch, books to read, travel destinations, and more. As long as she can create and check off items from her lists, life is good.

Unfortunately, not all is well today. Her favorite to-do app, the place where she stores all of her lists, behaves poorly when offline. When the Internet connection goes down or becomes unreliable, Susan has the following problems:

- Her to-do lists fail to load.
- She doesn't have confidence that the to-do items she created on other devices are being shown on this one.
- The source of the truth is in the cloud. When she is offline, her devices tend to diverge from this source in reconcilable ways.
- Susan attaches photos and videos to her to-do items. Any data connection slower than 4G is an exercise in patience.
- She has been an evangelist for this app, which has gotten some of her friends to use it. They like sharing items as a group, but when their phones are offline, they can't.
- The app has a tendency to show scary error messages. They make her think that she's lost some data. Sometimes she has, but it's never easy to tell.

Susan closed her laptop lid. The conference's Wi-Fi was really bad this year. Susan wasn't the kind to give up easily, but you can't be productive on an unreliable Internet connection. Fifteen disconnections in one hour? Really? A few seconds ago, her to-do app crashed, losing the last fifteen minutes of work. Just great!

She opened her phone. **No Signal** showed prominently in the upper left corner. No luck there either. Bother.

This seems to happen all the time lately. Her rural home is in a dead zone. Her commuter bus spends twenty minutes out of every hour in tunnels and behind mountain ridges. She encounters infamously bad Wi-Fi at every other Airbnb accommodation. Even at work, her Internet connection seems to go down once a week — and always at the worst times.

It wouldn't be so bad if my applications handled these situations gracefully, she thought; but they don't. They crash, and stutter, and freeze, and lose my data, and raise my blood pressure.

To be fair, not every application is that bad. E-mail is generally solid, e-readers have all the content that I want, already cached, and my office suite doesn't care at all. However, this to-do app drives me crazy, and it's not the only app that is unreliable. Social apps, mapping apps, audio streaming apps — they all have the same problems.

A thought occurred to her.

What if a proper offline experience was the starting point? What if developers made offline functionality their top priority instead of the first feature to get cut? What if we lived in a world that was optimistic about fast, reliable, and ubiquitous Internet connectivity but didn't assume its existence? People would find their apps more useful and have less frustration in their day-to-day tasks.

Phone and laptop forgotten for now, Susan pulled out her sketch pad and a black marker pen. She started drawing.

The offline paradigm

Internet connectivity is only one part of the puzzle. Modern devices rely on different types of wireless networks: GPS, LTE, 4G, 3G, Edge, GPRS, Wi-Fi, Bluetooth, and RFID, among others. If any of these go down, she mused, certain bad things can happen:

- If GPS goes down, it can mess up your fitness tracking resulting in erratic maps and false progress toward your fitness goals.
- As your mobile connection degrades from 4G to GPRS, web browsing slows and eventually stops. Wi-Fi has the same problem.

- Bluetooth is also important. It connects us to our keyboards, mouses, headsets, and other devices. As the signal weakens, these accessories become unreliable or cease functioning altogether.

- NFC, while not ubiquitous today, certainly will be so in the future. We'll depend on it for all or most of our **Point of Sale (POS)** financial transactions. If it goes down, can we use other networks as backup?

Susan stopped writing and put down her pen. She gazed reflectively at the list. The paper lay on top of her laptop as though mocking the brick that lay underneath.

So, with these failure cases, how might we develop our apps differently? More importantly, is there a common set of principles that we can use in order to avoid reinventing the wheel each time? What makes an excellent offline experience?

Hand reached for pen and she began a second list:

Principles for a great offline experience are as follows:

- Give me uninterrupted access to the content I care about.

- Content is mutable. Don't let my online/offline status change that.

- Error messages should not leave me guessing or unnecessarily worried.

- Don't let me start something I can't finish.

- An app should never contradict itself. If a conflict exists, be honest about it.

- When the laws of physics prevail, choose breadth over depth when caching.

- Empty states should tell me what to do next in a delightful way.

- Don't make me remember what I was doing last. Remember for me.

- Degrade gracefully as the network degrades. Don't be jarring.

- Never purge the cache unless I demand it.

Susan put her pen down. A good list. She gave a contented nod, put the paper in her bag, and walked out of the café, humming to herself. It was an interesting thought exercise and possibly more. She would bring it up with her design team at their 10 o'clock review. Their project might benefit from these principles.

Developing for a worst-case scenario

A common practice is to design for the happy path, a scenario where everything works correctly and according to plan. These scenarios tend to paint an overly optimistic picture of the real world. When the happy path meets real-world problems, the charade is over and the app must undergo serious revision or a rewrite.

This is bad for a number of reasons. Obviously, rewriting an app is costly. One of the tenets of good software development is to fail quickly and rapidly. If you finish your app, only to find that it doesn't work as advertised, you've failed far too late. Even if you don't have to rewrite the app, bolting on fixes to address fundamental problems with the underlying architecture can cause bugs that tend to linger until the inevitable rewrite.

The alternative is to start by designing for a worst-case scenario. Worst-case scenarios are more representative of the real world so apps are more likely to behave as you expect. These scenarios tend to be harder to solve but result in a robust solution that covers more edge cases. As a worst-case scenario is a superset of the happy path, you get two solutions for the price of one.

This strategy can be employed in many other areas, such as security, responsive web design, accessibility, and performance. The approach is simple. Pick one or more worst-case scenarios that impact your app the most and build with these scenarios in mind. A worst-case scenario is an excellent constraint. The more constraints you add, the easier it is to design a great solution.

So, what is an offline user experience? We can boil down the principles from Susan's list in the previous section to form a concise definition:

An offline user experience bridges the gap between online and offline contexts by providing equivalent functionality and content to the extent that technology and the laws of physics will allow.

To design this experience, it's vital to understand the pain and frustration that people encounter every day. There are a million ways to build an offline-first experience, but the correct solution is formed by the needs of your audience. The story in the previous section was one example but you can find an infinite number of scenarios. Once you know who you're developing for and their pain points, devising a solution is much simpler.

Let's go through a few more offline scenarios to get you going. Once you start thinking about who your audience is, write scenarios such as these to describe how the offline experience impacts them. Talk to people. Observe them in the real world. Do research to see what others have found. These are all the things that can help you build a more accurate picture of your target audience.

Going on an off-the-grid vacation

Once a year, Carl disappears into the wilderness for a two-week trek along the Pacific Crest Trail in the Western United States. He does this to get away from technology. However, sometimes, due to the nature of his job, he has to respond to an important e-mail or take the occasional phone call.

There is no cellular coverage out in the pines. To connect, Carl has to leave the forests behind and hitch a ride to the nearest hotel. After a hot shower and some real food, he powers up his Android and endures shoddy hotel Wi-Fi for the evening. It's an inconvenience but a compromise that Carl is willing to accept.

Living in a third-world country

Marsha owns an old Nokia phone with an i9 keypad. She has access to a slow GPRS network, which often goes down. As modern websites are so bandwidth-intensive, her phone so slow, and online connectivity so tenuous, she turns off images and media when she browses.

In the next town, the mobile network is slightly better but it's a five-mile walk. When Marsha needs to e-mail photos, she composes draft e-mails and queues them on her phone. Once a week, she treks into town to deliver milk from their small herd of cattle. During this time, she is able to send those e-mails. It's a slow and frustrating experience but she can't do much about it.

Commuting on public transportation

Elizabeth travels on a bus for an hour each way on her daily commute. The bus doesn't have Wi-Fi and encounters several dead zones on the route. She is a dedicated bibliophile, so these routes are a great opportunity to feed her reading addiction. As she reads 1-3 books a week, she often runs out of reading material.

When this happens on the bus, she can't usually do anything about it. Her life is so busy that she mostly doesn't have time to think about downloading new books before her commute. Maybe she'll bring a paperback along next time for something to do.

Working on a Wi-Fi only device

Shawon is 14. On his last birthday, his parents bought him an iPod touch. He wanted an iPhone for the coolness factor, but, oh well. It's fine. It doesn't have GPS or a mobile plan but most of the time, it's as good as a real iPhone.

Like most kids, he relies on his parents for transportation and accompanies them on their errands. As he has a Wi-Fi only device, he often goes somewhere new, only to have to figure out what the Wi-Fi password is—if the place even has Wi-Fi. There aren't many Wi-Fi networks without passwords these days.

When he can't find the password, he has to stay contented with the games and music that are on his device. However, most of his games require an Internet connection, which is odd, particularly for games with a single-player mode.

Sharing files between mobile devices

Francine likes her world where file sharing is so easy and seamless. In 2005, you used USB drives. In 2015, you share files wirelessly, with technology and services such as Dropbox, Google Drive, and Airdrop. Unfortunately, when her Internet connection dies, she's transported back to 2005.

Just last week, her editor asked for an update to an outline that she wrote. Traveling at the time, she didn't have easy access to Wi-Fi. She made the changes on her laptop but couldn't transfer the file to her phone in order to be e-mailed. After struggling with this for several minutes, her Airbnb host replied with instructions to connect to the network. Crisis averted.

Streaming a high-definition video from YouTube

Brian streams online videos, a lot. Most of this streaming is over his phone. His mobile provider is kind of horrible, so he has to deal with dodgy Internet connectivity. Most of the time, YouTube is very good about decreasing the quality of the videos to compensate for bad network conditions, but it isn't always enough.

He wishes there was a way to get a transcription of the video or just wait for the entire thing to download before playing it. YouTube won't download an entire video when it's paused. Unfortunately, these features don't exist. Instead, Brian bookmarks the videos to watch at home.

Online shopping with your phone

Jasmine is not an organized person. She likes taking care of things in the moment, as they come to mind. As a busy professional, she doesn't have the mental resources to memorize lists and doesn't want to be tied to a to-do app or sheet of paper.

Shopping is one example. Throughout her day, she does a mental inventory as she moves about. When she spots an item in short supply, she grabs her phone and places an order with her Amazon app. When she's connected to the Internet, this works great. When she's not, it frustrates her that she can't take care of things. If only Amazon provided a better add-to-cart experience for people who want to buy things while offline.

Getting work done at a conference

Conferences are notorious for their abysmal Wi-Fi. Raphael is attending Whole Design, along with 1,500 other attendees. It's a three-day, single-track conference. On day two, with four hours to go, he is ready to be done fighting for bandwidth. He hasn't gotten much done beyond seeing some great talks. Even his cell phone, normally reliable, doesn't work here at all.

Every couple of hours, the attendees get a 30-minute break. Raphael uses these opportunities to sprint outside the venue. Here, he can get a solid 3G connection. This is enough to check his e-mail and chats and generally check in with his world. At least, it should be. His apps behave as though there's a binary switch between no bandwidth at all and enough to stream an HD video. A third option would be nice.

Principles of a good offline design

Once you've compiled a few scenarios such as these, identify some common principles to build your app around. Here are the principles that Susan identified at the beginning of the chapter, with a more detailed explanation for each one.

Give me uninterrupted access to the content I care about.

When offline, you don't want to open your app and have no content to consume or interact with. You want to have lots of content available. As an extreme example, a web browser that allows you to surf the entire Internet, as it appeared when you were last online, would be a fantastic experience. This doesn't exist because the laws of physics prevent such an experience.

Naturally, you can't provide an infinite amount of content for people. Instead, provide as much content as possible without hurting the user experience in other ways (that is, don't steal all their bandwidth, don't fill up all the space on their device, don't let the performance suffer, and so on).

Content is mutable. Don't let my online/offline status change that.

One common practice is to make content read-only while offline. When online, you can edit, delete, or manipulate the content, but offline, the only thing that you can do is view it. People want to work with their data regardless, but you're telling them that they can't, usually for no good reason.

Instead, as much as possible, make online behavior match offline behavior. Once you've decided what people can do with their data, try to apply these decisions to all situations. You'll make their lives easier, and yours, as you only have to design and build a single interaction paradigm.

Error messages should not leave me guessing or unnecessarily worried.

A cryptic error message is a bad error message. Well-designed applications should be clear, using human language and a consistent vocabulary. Error messages should be no different. Unfortunately, error messages are often overlooked in the design process. When this happens, the application assumes two faces: a friendly face when everything is fine and a distressing face when things go wrong.

The effect can be jarring and incomprehensible to some people. When their world is falling apart, they need more reassurance and help from their tools, not less. Make sure that you think about the wording in the error messages and how errors present themselves visually.

Don't let me start something I can't finish.

While performing a task, the further I get, the more vested I become in the outcome. For example, let's say I start a task consisting of eight steps. On step seven, if you tell me that I can't finish because I'm offline, I'll be frustrated. If you tell me that I can't save my progress and will have to try again later, I'll be furious.

Don't do that. Be up front about what I can or cannot do. Don't let me proceed down a path with a known dead end.

An app should never contradict itself. If a conflict exists, be honest about it.

In *Chapter 6, Be Eventually Consistent* we'll discuss split-brain scenarios. A split-brain scenario occurs when two people change the same data at the same time, on different devices. When the devices try to reconcile their data automatically, they can't. It's up to people to decide what change to keep and what to throw away.

In a poorly designed application, several things can happen. The app can clobber one of the changes without asking, resulting in confusion. It can show both of the changes side by side without explanation, which is also confusing. It might also enter a cryptic failure state, which leaves people even more confused and powerless.

Always get input from people when a conflict cannot be automatically resolved. If you must show inconsistent data, always provide an explanation.

When the laws of physics prevail, choose breadth over depth when caching.

As mentioned in the first principle, infinite caching violates the laws of physics. There will often be more data than you can reasonably cache. You have to make a prioritization decision. What data is most valuable to people? Cache that data first.

In most applications, there are different types of data that can be cached. As older data is usually less valuable than newer data, the more you cache, the value of caching in a particular category goes down. At some point, continuing to cache a particular data type is less valuable than switching to a different data type.

First, cache a small set of data across all the types. Then, go progressively deeper across all the data types simultaneously, with the more valuable types prioritized. As a result, no empty screens will appear when people skim the app. The UI will feel data rich on the surface and deep in all the right parts, which is what they want.

Empty states should tell me what to do next in a delightful way.

If there is nothing in the cache, you will have to show an empty state. There is very little that people can do with these states, which is a bad experience.

However, these states can be more than blank screens. They can tell people what to do next. They can be delightful: providing a clever message, image, animation, or video that people wouldn't ordinarily see. They can contain sample data, which people can interact with to see how the screen would ordinarily work. They can even keep people entertained with a game, puzzle, or other distractions.

Empty states are very easy to ignore. When engineering resources are tight, they are among the first things to be cut. Figure out how often people will see these relative to the other screens and then apply resources appropriately. During offline usage, empty states are more likely to occur.

Don't make me remember what I was doing last. Remember for me.

Context is king. If I close my app, preserve my current state. When I reopen the app, I want to see what I was doing last or find it simple to get there. The more complicated the app, the more important this is.

Many apps forget the context when switching from online to offline. They give themselves permission to reset certain parts of the UI, which is jarring and easily avoidable. As with most of these other principles, don't compromise on the attributes of a good experience just because an app happens to be offline.

Degrade gracefully as the network degrades. Don't be jarring.

Networks are not binary. There is a lot of gray area between not working at all and working very well. One of the goals of an invisible design is to avoid speed bumps in the user experience. An interface may switch abruptly from an online mode to an offline one, or worse, toggle rapidly between the two states. If implemented poorly, this will interrupt the flow and make it more difficult for people to get their work done.

Help the application see or anticipate an impending loss of connectivity. When appropriate, it should scale back its bandwidth usage, avoid massive changes to the interface, and gently nudge people if they are encountering any problems. The goal is a smooth transition from online to offline that informs the user in an unobtrusive way.

Never purge the cache unless I demand it.

Cache is the most valuable aspect of an offline experience. If the cache goes away while offline, people won't have any content. This is a bad experience. Protect the cache at all costs. Don't clear it unless people ask for it explicitly (usually for security reasons).

The latter implies an affordance to clear the cache. Web browsers have one but a lot of apps don't. This is extremely important, particularly when login credentials are involved. People may want to purge their data for a good many reasons. They may even need to do it remotely from another device.

Making the case to yourself

Unreliable and nonexistent Internet connectivity is a widespread problem, both in highly developed nations and nations that are less developed. Consider a few statistics from ITU, the United Nations specialized agency for information and communication technologies (`http://www.itu.int/en/ITU-D/Statistics/Documents/facts/ICTFactsFigures2015.pdf`):

- As of 2015, 4 billion people from developing countries remain offline
- Only 9.5% of the people in the **least developed countries** (**LDCs**) use the Internet
- Only 29% of the people in rural areas have mobile coverage that is 3G or faster
- Only 1 in 100 people in Africa have a fixed broadband subscription

If you optimize for a fast Internet connection, you are making an implicit decision about who you're designing for. If you don't want Internet connectivity to dictate who can use your app, design the offline experience first. Even if you're building an application that only Americans will use, consider that potentially 3 out of every 100 people are eking out a mere 56 kbps. Every megabyte has a large impact on their user experience and ISP bills.

This doesn't mean that you have to build an app that looks like **Craigslist**. There are intelligent ways to make an app scale its behavior based on the Internet connection available. You can even use location services to become aware about the potential dead zones in time to take preventative measures. We'll cover these strategies in *Chapter 8*, *Networking While Offline*.

Are you convinced that building with an offline-first mindset is a good idea? Great! Now, how do you sell this vision to the decision makers around you? Many activities have a negative impact on productivity in the short-term but a positive impact in the long-term: writing tests, conducting user research, including good comments in the source code, and so on. In the same way, switching to an offline-first paradigm can be perceived as an unnecessary distraction, even though the long-term benefits are clear. Let's address this now.

Making the case to others

If offline usage is so important, why don't we optimize for it? One reason is internal: it's hard to change our innate behaviors and habits. Developers are on the cutting edge and have difficulty imagining a world where computers are slow and Internet connections are flaky. Fortunately, there are tools to help us simulate that other world.

Even if we overcome this obstacle, there are still external hurdles to change:

- We have to convince other developers that this will help us build better products, more efficiently
- We have to convince other designers that this will improve the user experience
- We have to convince our boss or others in upper-level management that this will increase our revenue

These are harder conversations to have but worthwhile. One of the goals of this book is to provide the evidence that you need to convince the developers, designers, and management in your organization. Offline-first development can improve your ability to deliver in all of these areas.

Convincing other developers

Developers care about efficiency and don't want to write more code than necessary. The extra care that goes into an offline-first application can, at a glance, seem to necessitate more code. You might get pushback that the team wants to build the easiest thing first and leave the offline functionality for later. This is a mistake. Instead, have a discussion about the architecture and tools that can ease the development effort.

In *Chapter 4*, *Getting Online*, we'll show how tools such as **PouchDB**, **remotestorage. io**, and **Hoodie**, custom built for offline use, can greatly reduce the amount of code that your team will need to write. These tools allow you to write good offline code, once. As applications are built, they gain complexity. This complexity makes them more difficult to change in the future. Using the right tools up front can save days or weeks of development time over the long run.

To make the case, show all the ways that an online app that is not optimized for offline usage can fail. The following chapters will talk about these situations in detail. Describe the pain that users experience while encountering these scenarios. If possible, take an existing app and write a task-based script that exercises these scenarios.

If your team is still skeptical, find people willing to try offline tasks on the app that you've built or a simple prototype. Invite your coworkers to attend. There's nothing like a firsthand observation of people struggling with a key part of your app to convert team members to your side.

Convincing other designers

Functional designers care about how people solve problems using the app. The trick is that there are many problems to be solved and you can't expect others to give offline problems the priority that they deserve. One way to measure priority is by the number of people affected by the problem. As offline problems are so widespread, you can make a strong case that every person using your app will benefit to some extent.

To start the discussion, build a task flow diagram for your application. This will help you see the scope of the app from a workflow perspective. In the diagram, point out the offline error cases and how these will affect the experience that people have. When functional designers see all the dead ends that their target audience might experience, they will be motivated to join you.

Visual designers may be a bit harder to convince. They too care about the experience of an app but care more about the clarity of the presentation rather than the functionality. Take the task flow diagram that you built earlier and mock rough screens that correlate to each step in the flow, including the error cases. Make sure that everyone is aware of the total surface area to be designed, not just the happy path. Often, designers don't see the whole picture up front because they're working from a theoretical model of how things might work, not a model that has been battle-tested in the real world. By building an accurate map of the world, your designers become better equipped to design a clear and delightful interface.

Convincing your boss or the upper-level management

The business cares about generating revenue. This goal is often in opposition to spending time on quality software that works well in a variety of adverse conditions. Correspondingly, these are some of the hardest people to convince. You must show that changing the way you do design and development will positively impact the bottom line.

People have a very low tolerance for apps that don't function. Only 16% of those who fail to use an app the first time will return (`http://techcrunch.com/2013/03/12/users-have-low-tolerance-for-buggy-apps-only-16-will-try-a-failing-app-more-than-twice/`). If one of the 4 billion people that are without a regular Internet connection tries to use your app and can't because they're offline, how long will they stay with yours before uninstalling and switching to a competitor's app?

On the other hand, if you're the only app among your competitors to provide an exceptional offline experience, you're more likely to gain market share and positive reviews from the billions of people impacted by poor offline experiences.

Revenue aside, if there are people in the management chain who care about design, your job is easier. The higher they are in the chain, the easier still. You can argue with them about design on even terms as they're already convinced of its value. If the CEO wants well-designed products, your job as a design emissary is done.

Summary

In this chapter, we established a case for why you want to approach application development with an offline-first mindset. In the next chapter, you will build a basic offline to-do app. Each subsequent chapter will build on this base. By the end of the book, you will have transformed this base into a fully capable online app that translates seamlessly between the online and offline worlds.

Open your development environment and let's get started.

2
Building a To-do App

In *Chapter 1*, *The Pain of Being Offline*, we laid the foundation for why offline-first development is important. Now, let's build an app with this approach. The app that we'll be building is a to-do app that has been written using the **Sencha Touch JavaScript** framework. As Sencha Touch is just JavaScript under the hood, you can view it in any browser. Additionally, if you're a mobile developer, fear not. We'll use **Cordova** to deploy your app to an iOS device as well. The section on Cordova is entirely optional and works for Android as well, with minor modifications along the way.

Let's start by getting your development environment configured. For simplicity, I assume that you're running Mac OS X and Chrome. The instructions may differ slightly with different operating systems and browsers. I'll call out the differences as we go along.

Configuring the development tools

There are three tools that we will use in most of the examples: a text editor for the coding, a browser (preferably Chrome) to run and debug your app, and the command line to run the web server and execute commands. Let's walk through each of these tools.

A text editor

If you have an existing editor that you like, use that. In this book, I will use **Sublime Text 3**. You can download a free evaluation of Sublime for OS X, Windows, and Linux. If you prefer to use a free editor, good choices include Notepad++ (Windows), Bluefish Editor (OS X and Linux), and GVIM (Linux). To install Sublime, follow these instructions:

1. Open `www.sublimetext.com` in your browser.

2. Click on **Download**.

3. Click on **Sublime Text 3**.

4. Choose the appropriate version for your operating system.

5. Wait for the download to complete and run the installer.

You should be able to run Sublime now. It's a GUI-based editor that is very intuitive and simple to use. If you're interested in getting more out of Sublime, check out the documentation at `www.sublimetext.com/docs/3/`.

A browser

I recommend Chrome because it is cross-platform and has an excellent suite of developer tools, including tools for mobile development. You can **Download Chrome** from `www.google.com/chrome`.

Once Chrome is installed, start it. To open the developer tools, click on **View | Developer | Developer Tools**. Then, to enable the mobile app view, toggle the device mode:

Mobile development toggle

This gives you a grid and viewport sized to the dimensions of a mobile device. You may need to resize the browser and refresh the page for the changes to take effect:

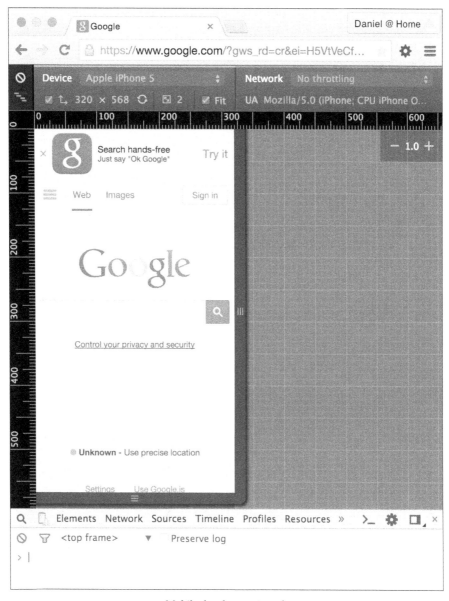

Mobile development mode

A command line

Use the default command-line tool that comes with your operating system. For OS X, that's a terminal. For Windows, that's cmd.exe. For Linux, that's (usually) bash. As we build the app, I'll provide most of the commands that you need to run, but here are a few additional commands that you may find helpful when navigating and manipulating the filesystem:

Command	Description
ls (dir on Windows)	This will list the contents of the current directory
cd [./|../|/] (cd [.|..|C:\] on Windows)	This is the change to a directory starting from the current/parent/root directory
mkdir dir_name	This will create a directory named dir_name
⌘ + C (*Ctrl* + C on Windows)	This will terminate the currently running command

Homebrew

There are a couple of non-native package managers that you will need later in development. As OS X does not include a native package manager, Homebrew was created to service this need. Let's install it now:

1. Open www.brew.sh.
2. Copy the Ruby command:

Copy the Homebrew install command

3. Paste the command in your terminal:

Install Homebrew

4. Follow the installation instructions.

    Now, when you need to install a package with Homebrew, use the following syntax on the command line:

    ```
    $ brew install <package>
    ```

npm

What Homebrew is for OS X, npm is for JavaScript. It comes preinstalled as a part of Node.js and as we'll need this library as well, let's install both now:

```
$ brew install node
```

Java

You will need a Java Runtime Environment to run some of the commands in this book. First, check to see if Java is already installed:

```
$ java -version
```

If Java is installed with version 1.7 or greater, you're ready to go. If not, install a recent version of Java:

1. Open www.java.com/en/download.
2. Click on **Free Java Download**.
3. Click on **Agree and Start Free Download**.
4. Once the download is complete, open the file that you downloaded.
5. Click on **Continue**.

6. Click on **Install**.

7. Enter your credentials, if requested.

8. After the installation is complete, click on **Close**.

Finally, verify that Java and its proper version has been installed:

```
$ java -version
```

Arranging your workspace

Almost as important as the tools themselves is the way you arrange them on your screen. The exact layout may vary depending on the size and number of your screens, but on a 15-inch laptop, the following setup works well for me:

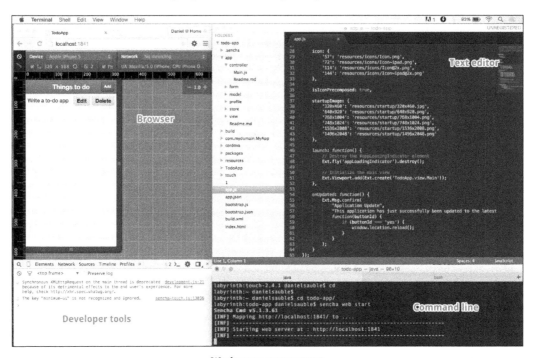

Workspace arrangement

On the bottom right-hand side is my command-line tool. I have two tabs that are open. The first tab is where I run my web server. The second tab is where I input other commands and is used more frequently of the two.

On the top right-hand side is my text editor, where I can browse the project files and edit code. When I make a change, I use ⌘ + *Tab* (*Win* + *Tab* on Windows) to switch to my browser and preview the change.

On the left-hand side is my browser. I use this to preview changes to the code, monitor the developer console for error and debug messages, and inspect the DOM if I'm trying to troubleshoot a bug.

Additionally, not pictured, I have a separate fullscreen browser in order to browse documentation and search Google for help when I run into problems. If you have a second screen, it would be a great place to put this.

Now that you have your development tools and workspace configured, let's get the toolchain up and running.

Configuring Sencha Touch

Sencha Touch is a JavaScript-based MVC framework used to build cross-platform mobile apps. It's a proprietary framework but free for open source projects, so we'll use it for our to-do app. To get and configure it, follow these instructions:

1. First we will download Sencha Touch:
    1. Open `https://www.sencha.com/products/touch/` in your browser.
    2. Click on **DOWNLOAD FOR FREE**.
    3. Fill out the contact form and click on **DOWNLOAD TOUCH**.
    4. Check your e-mail for a link to download Sencha Touch.
    5. In the e-mail, click on **Download Sencha Touch**.
    6. After the download is complete, extract the ZIP in its own folder.
    7. Open this folder (it should be named `touch-2.4.1` or similar).

2. Install Sencha Cmd:
    1. Open `https://www.sencha.com/products/extjs/cmd-download/` in your browser.
    2. Click on **DOWNLOAD WITHOUT JRE INCLUDED**.
    3. Once the download is complete, run the setup application.
    4. Accept all defaults and click on **Install**.
    5. Once the installation is complete, click on **Finish**.

3. Install Ruby:
    1. **Mac OS X**: Ruby is preinstalled. Verify this by running `ruby -v` on the command line.

2. **Windows**: Open `http://rubyinstaller.org` in your browser. Click on **Download**. Choose a **RubyInstallers** for Ruby 2.1.6 or greater and click to download. Run the installer after the download is complete.

3. **Ubuntu**: Run `sudo apt-get install ruby2.0.0` to download and install Ruby. If on a different Linux distribution, use whatever native package manager comes with your operating system.

4. Start the web server:

   1. Start the web server by running `sencha web start` on the command line.

   2. Open `http://localhost:1841` in your web browser.

   3. If the installation has been completed successfully, you should see a page as follows:

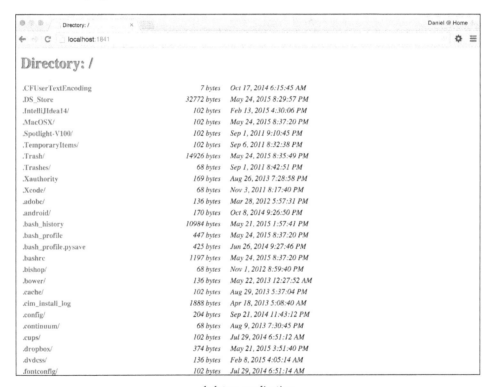

skeleton_application

   4. Stop the web server by typing ⌘ + C in the command terminal window.

Now, Sencha Touch is running successfully on your system. Great! Next, we'll prepare a directory for your application, start developing the application itself, and commit it safely to GitHub.

Creating a Git repository

This step is optional but highly recommended. It will prevent you from losing progress when you accidentally delete files and allow you to easily revert changes or try experiments without fear of diverging from the examples in the book.

You may also choose to use GitHub (or another code sharing site) to upload your code. This protects your code from hard drive failure and makes it easier to share and collaborate with others. Again, this step is optional but recommended. If you want to use Git and GitHub, perform the following steps:

First, we will install GitHub. The following instructions are for OS X only. If you're on Windows, download the GitHub client from `windows.github.com` and install it the way you would install other Windows applications:

1. Open `mac.github.com` in your browser.
2. Click on **Download GitHub Desktop**.
3. After the download is complete, open the application.
4. Click on **Move to my application folder**.
5. Click on **Continue**.
6. If you don't have a GitHub account already, create one here; otherwise, log in with your existing credentials.
7. Click on **Continue**.
8. Enter your name and e-mail address.
9. Click on **Install Command Line Tools**.
10. Click on **Continue**.

11. Click on **Done**:

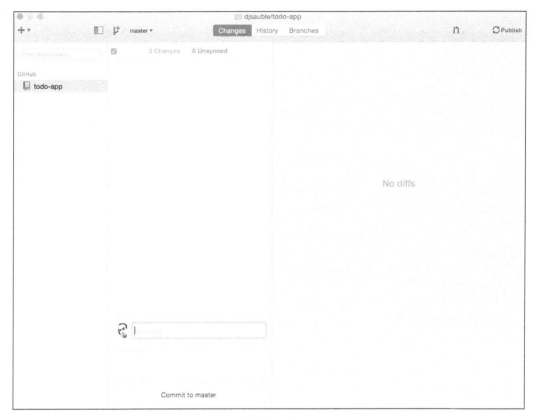

GitHub tool

Create a repository and upload it to GitHub:

1. Click the + icon to create a repository.
2. Type todo-app for the repository name.
3. Click on **Choose...**.
4. Select a folder on your computer where you want to develop the app. (I recommend creating a projects directory under $HOME.)
5. Click on **Open**.
6. Click on **Create Repository**.
7. Click on **Publish** to upload your new repository to GitHub.
8. Type Initial commit in the description field.
9. Click on **Push Repository**.

That's it! After you publish to GitHub, you can log in to your GitHub account and view these changes from any computer or device. You can download the repository, share it with others, and even make changes directly through your browser.

GitHub is free for **open source software** (**OSS**) projects. If you want to make the repository private, GitHub charges a reasonable monthly fee. Alternatively, you can choose not to use GitHub, but you'll want to make sure to back up your repository.

As we continue to develop the app, we'll prompt you to commit your changes to Git. As a developer, committing to Git should be as common and natural as saving files in your editor.

Designing the offline experience

In the last section, you configured the tools necessary to make your app real. Now, let's step back and design the app in the context of some of the offline scenarios that we discussed in *Chapter 1*, *The Pain of Being Offline*.

What do users need from this app and why?

Designers often employ user stories to break a scenario into chunks. Each user story takes a person, describes a specific need that they have, and reveals the context behind the need. The format that we will use in this book is as follows:

- As [a person]
- I want [to accomplish a goal]
- So that [context about the goal]

Let's talk briefly about each clause. How do user stories help us understand what we need to design and build?

- **As [a user]**: This helps us understand who we're building for. If this person works with other people, we can (and should) consider how they will work together as a team. This is extremely important in communication-oriented products. Our simple to-do app considers a single user, Susan, but you could easily write additional stories that reference her friends, acquaintances, and coworkers.

- **I want [to accomplish a goal]**: This helps us understand what the person is trying to get done. It should focus on actions and goals with a concrete outcome. By itself, this isn't enough to tell us what to build. We still lack the context around the need. That's what the next clause informs us about.

- **So that [context about the goal]**: Unlike the *As* and *I want* clauses, there may be multiple *So that* clauses in a single story. The context behind a need may be complex so we should capture it fully. These help us understand the motivation behind the need. Motivation is important because the same need with different motivations can manifest itself in very different solutions. For example, *so that I don't lose my job* will result in a very cautious, accident-averse solution as compared with *so that I learn to fail quickly*.

Once you've written one user story, write another. Depending on the complexity of the scenarios and granularity of each story, an app may have as few as five user stories or as many as fifty. However, it is unlikely that you will build all of them at once. This is an exploratory phase to discover needs that are both primary and ancillary to the app that you're building.

User stories do not live in isolation. If your app is at all complex, you should map the stories relative to each other. This will help you spot patterns that may influence the way you choose to implement them. Give each user story a brief title in order to make it stand out at a glance, then group the stories by user, function, or complexity. We won't map our to-do app stories as there are so few of them, but definitely read about story mapping online if you want to learn more.

Here are five user stories associated with Susan's scenario from *Chapter 1, The Pain of Being Offline*:

- **Items should display while offline**: As Susan, I want my to-do items to always appear once they are downloaded so that I can remain productive while offline, I don't have to worry about data loss, and I can learn to trust my to-do app to *Do The Right Thing™*.

- **Keep all my devices in sync**: As Susan, I want the to-do items on all my devices to remain in sync so that I don't have to wonder whether a device has a particular to-do item, I can just grab the nearest device and start typing, and my source of truth is the list itself, not one of several devices.

- **Don't keep me waiting**: As Susan, I want my to-do app to feel fast and ready to go at all times so that editing my to-do list is as frictionless as pen on paper, I never feel like I'm waiting, and I don't give up and neglect to write something down.

- **Sharing with friends should be bulletproof**: As Susan, I want to share items easily with friends regardless of where we are so that we can coordinate ourselves, sharing with friends who are close by feels visceral and physical, and I don't have to rely on third-party solutions, hacks, or intermediaries.

- **Safeguard my data**: As Susan, I want everything that I type in the app to be preserved so that I don't lose anything accidentally, the app earns my trust, and I can always go back, revert, undo, and so on and so forth.

How will people accomplish their goals using the app?

User stories do two things: they point us in the direction of the final solution and give us constraints that eliminate options. Both of these functions are valuable but still very abstract. How do we flesh out the design in more detail?

One excellent approach is through workflows. Workflows describe a path, or paths, through an application by which people can accomplish their goals (as illustrated in the user stories). They are most often visualized as flowcharts.

The following flowchart describes the offline workflows through our to-do app. The chart consists of two types of components: rectangles and circles. Rectangles represent specific views in the application by which people can observe the state of the UI and orient themselves to it. Circles represent specific actions that people may perform on a view:

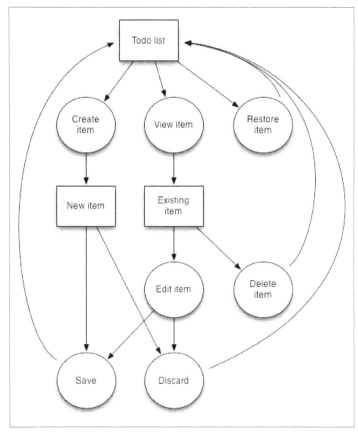

Workflow

With this basic workflow in place, let's look at each user story and see how the workflow maps to it. Some of the user stories imply an online experience and thus won't be solved in the first offline version of the app. We'll note this where appropriate.

Items should display while offline

As the entire app is offline, this is a trivial problem to solve. Items created in the to-do app are always stored locally and thus always available for display. We will solve this more rigorously in *Chapter 4, Getting Online* for the online case.

Keep all my devices in sync

As the first version of the app is offline, each instance of the app has its own local database with no syncing required. In *Chapter 4, Getting Online*, we will introduce multidevice support and add incremental improvements in *Chapter 5, Be Honest about What's Happening*, *Chapter 6, Be Eventually Consistent*, and *Chapter 8, Networking While Offline*.

Don't keep me waiting

Again, this is trivial to solve in an offline context. Loading data from the local storage is extremely fast. When we add online support, we will have to account for network bandwidth and latency with delays of milliseconds (or even seconds). We will address these concerns in *Chapter 7, Choosing Intelligent Defaults* and *Chapter 8, Networking While Offline*.

From an interaction perspective, the workflow implies a maximum of three clicks for any path. The worst-case scenario—editing an item—involves the following steps:

- Click an item in the list to view it
- Click to edit the item
- Click to save the item

This is reasonable but could be improved to one or two clicks if we employed inline editing or autosave. For the sake of simplicity, I will leave this as an exercise for the reader.

Sharing with friends should be bulletproof

Technically, sharing does not require an Internet connection. With Wi-Fi and Bluetooth, it is possible to share to-do items without an Internet connection. We will solve basic sharing over the Internet in *Chapter 4, Getting Online* with an exploration of Wi-Fi and Bluetooth in *Chapter 8, Networking While Offline*.

Safeguard my data

While offline, the phone is the source of truth for the data. As long as the phone has regular backups, data is relatively safe. However, with an active Internet connection, we can back up the data to the cloud at more frequent intervals. We will implement a basic Internet backup solution in *Chapter 4, Getting Online* with improvements in *Chapter 5, Be Honest about What's Happening* and *Chapter 6, Be Eventually Consistent*.

What does the app look like, roughly?

In the workflow, we can see that there are three primary views in the app: a list of to-dos, a form to create a new item, and a form to edit an existing item. Let's build these with some basic wireframes.

It's often the case that you don't end up with a carbon copy of your initial design. As you iterate on the design and turn it into code, you will discover obstacles, easier ways to do something, or a better approach. As long as the finished solution works and complies with your requirements in a well-designed way, it's actually better to adjust your design to these realities as you go along.

As we build our to-do app, you'll notice some of this behavior. In this chapter, compare our wireframes with the finished product. While we won't go into the reasons behind the discrepancies, think about why these differences exist. Would it have been better to stick with the original design or is the final app actually simpler and better than it otherwise would have been?

A list of to-dos

This screen displays a list of outstanding to-do items. It allows you to create new items, view the existing items, and restore the **Deleted items**. Here is one possible solution:

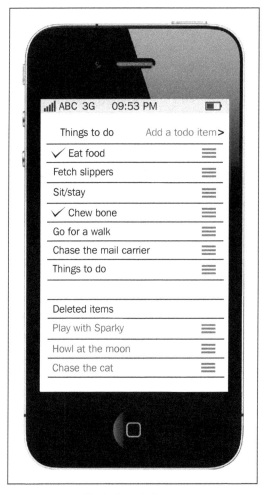

To-do list wireframe

There are two lists. The top list contains all the active items and the bottom list contains all the inactive items. To create a new item, click **add todo item**. To view an existing item, click it. To restore a **Deleted items** (or delete an active item), drag it from the bottom list to the top list (or from the top list to the bottom list).

Creating a new item

This screen is used to add a new item to your to-do list. As the workflow doesn't provide any insight into what a to-do item consists of, what are Susan's use cases? She wants her to-do app to be quick and easy to interact with, so we'll add as few fields as possible:

- **Text description**. This is how Susan refers to her to-do items. The description should be short and concise so a simple input field will suffice.

- **Image**: Susan is a visually-oriented person. She takes pictures of things to get done on her iPhone and then tags her to-do items with the images.

- **Location**: She doesn't want to have to check her to-do list all the time. Instead, she wants her phone to notify her when she's in the same location as a task that needs to be done.

With these criteria in hand, here's one possible solution for this screen:

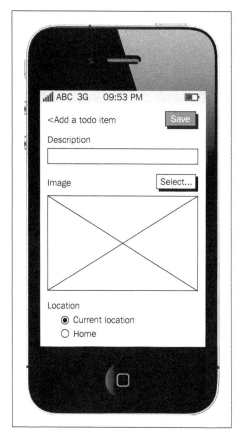

New item wireframe

Click the **Description** field to set a description. Click on **Select...** to attach an **Image** or video to the item. Choose a radio button to set a **Location** for reminders. Click on **Save** to commit your changes or click **Add a todo item** to go back without saving.

Editing an existing item

This screen is very similar to the new item screen. It is used to make changes to an existing to-do item, which means that the design is identical, except that the fields are prepopulated by whatever is assigned to the to-do item:

Edit item wireframe

Creating a skeleton application

Now, let's switch from design to code and create a skeleton Sencha Touch application. This will show us that the framework is functional and give us a base to start converting our workflows and wireframes into code.

Generating files with Sencha cmd

Using what you learned from setting up your development environment, switch to the `todo-app` directory that you used to create a repository. Now, generate a new application in this directory:

```
$ sencha -sdk /path/to/sdk/touch-2.4.1/ generate app TodoApp .
```

This will create a skeleton application with all the files needed. Now, verify that everything works as expected. You'll want two command windows open: one for the web server and the other for any commands that you need to run. In the first tab, start the web server:

```
$ sencha web start
```

In the second tab, preview the page in your browser:

```
$ open http://localhost:1841
```

If everything worked correctly, you should see the following sample app in your browser. You may need to toggle the device mode first.

Welcome app

Use the GitHub client to commit your latest changes. Give the commit a summary (and description, if you like), then click on **Commit to master**.

If you click the **History** tab, you can see your first commit. Click on **Publish** to upload this commit to GitHub as well. As we proceed, we'll prompt you to commit after completing a change. In general, it's a good habit to commit small chunks of code at frequent intervals as this makes debugging easier (and reduces the risk of lost work).

Creating the main view

Now, let's rip out the sample views and build our own. These views won't be very interactive but enough to show the rough shape of the final app.

In Sublime, open the `todo-app` folder. This will expose all the files under this folder for easy access in Sublime.

Now, open `todo-app/app/view/Main.js`. This file contains the code that describes the views in the skeleton application. Replace the existing code with the following:

```
Ext.define('TodoApp.view.Main', {
    extend: 'Ext.tab.Panel',
    xtype: 'main',
    requires: [
      'Ext.TitleBar'
    ],
    config: {
      tabBarPosition: 'bottom',

      items: [
        {
          title: 'List',
          iconCls: 'home'
        },
        {
          title: 'Create',
          iconCls: 'action'
        },
        {
          title: 'Edit',
          iconCls: 'action'
        }
      ]
    }
});
```

Refresh your browser. This code created three empty panels named **List**, **Create**, and
Edit, which you can open by clicking their respective icons in the toolbar. The toolbar
is temporary until we wire these views to each other directly:

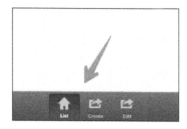

View switcher

Adding the list view

Now, let's recreate the wireframes by adding controls to each panel in turn. First, edit
the list panel to include the following code:

```
{
    title: 'List',
    iconCls: 'home',

    xtype: 'list',

    store: {
      fields: ['name'],
      data: [
        {name: 'Eat food'},
        {name: 'Fetch slippers'},
        {name: 'Sit/stay'},
        {name: 'Chew bone'},
        {name: 'Go for a walk'},
        {name: 'Chase the mail carrier'},
        {name: ''},
        {name: 'Deleted items'},
        {name: 'Play with Sparky'},
```

```
            {name: 'Howl at the moon'},
            {name: 'Chase the cat'}
        ]
    },

    itemTpl: '{name}',

    items: [
      {
        docked: 'top',
        xtype: 'titlebar',
        title: 'Things to do',
        items: {
          align: 'right',
          text: 'Add'
        }
      }
    ]
  }
```

This should result in the following screen:

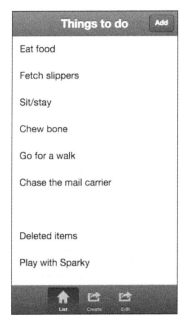

Mocked list screen

Adding the create view

Now, edit the create panel with the following code:

```
{
  title: 'Create',
  iconCls: 'action',

  items: [
    {
      docked: 'top',
      xtype: 'titlebar',
      title: 'Add item',
      items: {
        align: 'right',
        text: 'Save'
      }
    },
    {
      xtype: 'formpanel',
      height: '100%',
      scrollable: true,

      items: [
        {
          xtype: 'fieldset',
          title: 'Description',
        items: {
          xtype: 'textfield',
                  name: 'description'
          }
        },
        {
          xtype: 'fieldset',
          title: 'Image',
          items: {
            xtype: 'button',
            text: 'Select image…'
          }
        },
        {
          xtype: 'fieldset',
          title: 'Location',
          defaults: {
```

```
            labelAlign: 'right',
            labelWidth: '240px',
            xtype: 'radiofield',
            name: 'location'
                },
        items: [
          {value: 'here', label: 'Current location'},
          {value: 'home', label: 'Home'},
          {value: 'work', label: 'Work'}
        ]
      }
    ]
  }
 ]
}
```

This should result in the following screen:

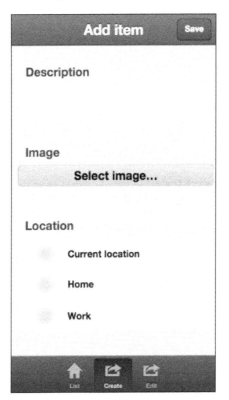

Mocked add screen

Adding the edit view

Finally, assign the following (nearly identical) code to the edit panel:

```
{
  title: 'Edit',
  iconCls: 'action',

  items: [
    {
      docked: 'top',
      xtype: 'titlebar',
      title: 'Edit item',
      items: {
        align: 'right',
        text: 'Save'
      }
    },
    {
      xtype: 'formpanel',
      height: '100%',
      scrollable: true,

      items: [
        {
          xtype: 'fieldset',
          title: 'Description',
          items: {
            xtype: 'textfield',
            name: 'description',
            value: 'Chase the mail carrier'
          }
        },
        {
          xtype: 'fieldset',
          title: 'Image',
          items: [
            {
              xtype: 'panel',
              width: '100%',
              height: 156,
              style: 'background: #AAAAAA'
            },
            {
              xtype: 'button',
              text: 'Remove image'
            }
```

```
              ]
            },
            {
              xtype: 'fieldset',
              title: 'Location',
              defaults: {
                labelAlign: 'right',
                labelWidth: '240px',
                xtype: 'radiofield',
                name: 'location'
              },
              items: [
                {value: 'home', label: 'Home', checked: true},
                {value: 'work', label: 'Work'},
                {value: 'other', label: 'Other'}
              ]
            }
          ]
        }
      ]
    }
```

This should result in the following screen:

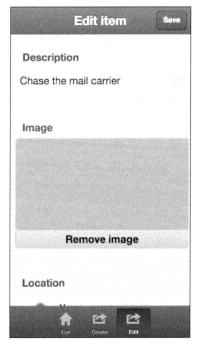

Mocked edit screen

When you start the application, you may notice a few warning messages in the developer console about missing requires. Add these requires to `Main.js` under the `xtype` attribute:

```
requires: [
  'Ext.TitleBar',
  'Ext.dataview.List',
  'Ext.form.Panel',
  'Ext.form.FieldSet',
  'Ext.field.Radio'
],
```

Committing your changes

At this point, you have mocked the wireframes in code. They don't do anything yet, but you can toggle through the three main views with the bottom tab bar.

 Now that the application is in a working state, commit these changes to Git before proceeding.

Building an offline to-do app

We cut some corners to put the wireframes in code. Let's get them the rest of the way there, make the pages functional, and set up the backing storage. After we've done this, we'll deploy the app to our phone and call it a day.

Breaking the views into their own files

Create three empty files in `todo-app/app/view/`:

- `List.js`
- `New.js`
- `Edit.js`

Now, break out the code for each tab in `Main.js` and put each into a separate file in the `todo-app/app/view/` directory with a few modifications.

First, take the contents of the innermost items array for the `List` view and put it in `List.js` in the following code block:

```
Ext.define('TodoApp.view.List', {
    extend: 'Ext.Panel',
    alias: 'widget.todo-list',

    config: {
    items: [
      {
        docked: 'top',
        xtype: 'titlebar',
        title: 'Things to do',
        items: {
          align: 'right',
          text: 'Add'
        }
      },
      {
...
      }
    ]
    }
});
```

Now, do this for the `New` and `Edit` views as well. Copy the fields to create a new item and drop them in `New.js` as follows:

```
Ext.define('TodoApp.view.New', {
  extend: 'Ext.Panel',
  alias: 'widget.todo-new',

  config: {
    items: [
      {
        docked: 'top',
        xtype: 'titlebar',
        title: 'Add item',
        items: [
          {
            align: 'left',
            text: 'Back'
          },
          {
            align: 'right',
```

```
              text: 'Create'
            }
          ]
        },
        {
  ...
        }
      ]
    }
});
```

Finally, copy the fields to edit an existing item and drop them in `Edit.js` as follows:

```
Ext.define('TodoApp.view.Edit', {
  extend: 'Ext.Panel',
  alias: 'widget.todo-edit',

  config: {
    items: [
      {
        docked: 'top',
        xtype: 'titlebar',
        title: 'Edit item',
        items: [
          {
            align: 'left',
            text: 'Back'
          },
          {
            align: 'right',
            text: 'Save'
          }
        ]
      },
      {

      }
    ]
  }
});
```

Now, convert `Main.js` to a regular panel view. As you only want to show one screen at a time, manually verify that the screens display as before by inserting `todo-list`, `todo-edit`, and `todo-new`, respectively, in the innermost `items` block:

```
Ext.define('TodoApp.view.Main', {
    extend: 'Ext.Panel',
    xtype: 'main',
    alias: 'todo-main',
    requires: [
        'Ext.TitleBar',
        'Ext.form.Panel',
        'Ext.form.FieldSet',
        'Ext.field.Radio',
        'Ext.List',
        'TodoApp.view.New',
        'TodoApp.view.Edit',
        'TodoApp.view.List'
    ],
    config: {
        layout: 'fit',
        items: { xtype: 'todo-list' }
    }
});
```

Does it work? Great! Check your changes in Git:

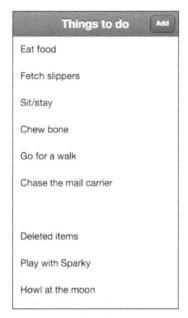

A list of to-do items

Connecting the views together

Note that we ditched the tab bar at the bottom of the screen. You won't use it to navigate, you'll click on the individual to-do items and buttons at the top of each screen. Let's wire these up now.

First, create a file under `todo-app/app/controller/` named `Main.js`. This is the main controller for the application. Paste the following code:

```
Ext.define('TodoApp.controller.Main', {
  extend: 'Ext.app.Controller',
  config: {
    views: [
      'List',
      'New',
      'Edit'
    ],
    refs: {
      mainPanel: 'todo-main',
      listPanel: {
        selector: 'todo-list',
        xtype: 'todo-list',
        autoCreate: true
      },
      newPanel: {
        selector: 'todo-new',
        xtype: 'todo-new',
        autoCreate: true
      },
      editPanel: {
        selector: 'todo-edit',
        xtype: 'todo-edit',
        autoCreate: true
      }
    },
    control: {
      'todo-list button[action=new]': {
        tap: 'showNewView'
      },
      'todo-list list': {
        itemtap: 'showEditView'
      },
      'todo-new button[action=save]': {
        tap: 'showListView'
      },
```

```
        'todo-edit button[action=save]': {
            tap: 'showListView'
        }
    }
},
showView: function(view) {
    this.getMainPanel().removeAll();
    this.getMainPanel().add(view);
},
showListView: function() {
    this.showView(this.getListPanel());
},
showNewView: function() {
    this.showView(this.getNewPanel());
},
showEditView: function() {
    this.showView(this.getEditPanel());
}
});
```

To make the application see this new controller, edit `todo-app/app.js` and add the following under the views array:

```
controllers: [
    'Main'
],
```

Notice how the controller identifies the buttons in each view by their action attribute. If you add multiple buttons in the same view and assign different actions, you can listen for their events with this method.

 Test the new changes by starting the web server and clicking through the application. Notice how the views change. Commit your changes to Git.

Creating a store to populate the list view

Now that you have the basic workflow in place, let's move the data from the view to a store. This will give us flexibility and persistence as the view will simply read from the store, which itself may be backed by any number of sources: an API, a database, or local storage.

As this app is offline, we must use a local form of storage. The easiest way to do this is with a browser's local storage. To accomplish this in Sencha, create a new model and store named `todo-app/app/model/Item.js` and `todo-app/app/store/Item.js`, respectively:

```
todo-app/app/model/Item.js:
Ext.define('TodoApp.model.Item', {
  extend: 'Ext.data.Model',
  requires: [
    'Ext.data.proxy.LocalStorage',
    'Ext.data.identifier.Uuid'
  ],
  config: {
    identifier: {
      type: 'uuid'
    },
    fields: [
      'id',
      'description',
      'media',
      'latitude',
      'longitude'
    ],
    proxy: {
      type: 'localstorage',
      id: 'todoapp-items'
    }
  }
});
```

The model defines the fields in each to-do item, specifies how unique IDs will be generated, and indicates that each to-do item is to be stored using the local storage. Now let's create the accompanying store for this model, `todo-app/app/store/Item.js` model:

```
Ext.define('TodoApp.store.Item', {
  extend: 'Ext.data.Store',
  config: {
    model: 'TodoApp.model.Item',
    autoLoad: true
  }
});
```

The store is used by views to point to the list of the to-do items. We could create a new store each time we needed one, but by defining it here, we avoid repeating ourselves. To make these new classes visible to the app, add references in `todo-app/app/controller/Main.js` below the views array:

```
models: [
  'Item'
],
stores: [
    'Item'
],
```

Now that we've created the model, let's wire it up to the `List` view.

Edit `todo-app/app/view/List.js`, where we will define the `list`, remove the existing `store` definition, and point to the `store` that we just created:

```
{
  xtype: 'list',
  height: '100%',
  store: 'Item',
  itemTpl: '{description}'
}
```

Now, reload the app. No items appear in the list because the store is empty, but we'll fix this next.

In `todo-app/app/controller/Main.js`, instead of just showing the list view when the **Create** button is clicked, retrieve the form values and create a new record. To do this, make a new `function` named `createTodoItem` and add it to the controller:

```
createTodoItem: function(button, e, eOpts) {
  var store = Ext.create('TodoApp.store.Item');

  store.add(this.getNewForm().getValues())
  store.sync();

  this.showView(this.getListPanel());
},
```

In the `control` hash, make the `todo-new button` point to this function instead of the `showListView` function:

```
'todo-new button[action=create]': {
  tap: 'createTodoItem'
},
```

Add a selector to the `refs` array so that we can retrieve the form with `getNewForm()`:

```
newForm: 'todo-new formpanel',
```

Due to a bug in Sencha Touch, the `autoLoad` option for stores doesn't work on list components. To avert this problem, add the following function to the `List` view:

```
initialize: function() {
  // Autoload appears to be broken for DataViews
  Ext.getStore('Item').load();

  this.callParent();
}
```

Refresh your app, create a new item, and notice how it appears in the list. Pretty easy, huh?

Adding buttons to each to-do item

The next step is to hook up the edit and delete workflows. As there are two of these, we have to create separate buttons for each list item. As `Ext.List` isn't flexible enough to accommodate this, we'll need to switch over to `Ext.dataview.DataView`. Change the type of the list to `dataview`, tell it to `useComponents`, and specify what the default component type should be:

```
{
  xtype: 'dataview',
  height: '100%',
  useComponents: true,
  defaultType: 'todolistitem',
  store: 'Item'
}
```

If you've been paying attention, you'll notice that we specified the `defaultType` attribute in our `DataView` component. This default type doesn't exist yet so we need to create it. Let's do this now. Create `todo-app/app/view/TodoListItem.js` and add the following lines:

```
Ext.define('TodoApp.view.TodoListItem', {
  extend: 'Ext.dataview.component.DataItem',
  alias: 'widget.todolistitem',
  requires: [
    'Ext.Label'
  ],

  config: {
    description: {
      flex: 1
    },
    edit: {
      text: 'Edit',
      action: 'edit',
      margin: '0 7px 0 0'
    },
    destroy: {
      text: 'Delete',
      action: 'delete'
    },
    padding: 7,
    layout: {
      type: 'hbox',
      align: 'center'
    },
    dataMap: {
      getDescription: {
        setHtml: 'description'
      },
      getEdit: {
        setData: 'id'
      },
      getDestroy: {
        setData: 'id'
      }
    }
  },

  applyDescription: function(config) {
```

```
    return Ext.factory(config, 'Ext.Label',
      this.getDescription());
  },
  updateDescription: function(newDescription, oldDescription) {
    if (newDescription) {
      this.add(newDescription);
    }
    if (oldDescription) {
      this.remove(oldDescription);
    }
  },

  applyEdit: function(config) {
    return Ext.factory(config, 'Ext.Button', this.getEdit());
  },
  updateEdit: function(newButton, oldButton) {
    if (newButton) {
      this.add(newButton);
    }
    if (oldButton) {
      this.remove(oldButton);
    }
  },

  applyDestroy: function(config) {
    return Ext.factory(config, 'Ext.Button', this.getDestroy());
  },
  updateDestroy: function(newButton, oldButton) {
    if (newButton) {
      this.add(newButton);
    }
    if (oldButton) {
      this.remove(oldButton);
    }
  }
});
```

This defines three components for each item: a label so that you can see what the to-do item is about and two buttons so that you can edit and delete the item. For simplicity, checking an item off your list and deleting an item you don't intend to complete is the same action.

Now, in `todo-app/app/controller/Main.js`, add a reference to this new `views` so that we can talk to it:

```
views: [
  'List',
  'New',
  'Edit',
  'TodoListItem'
],
```

Refresh your app and notice how the list works as before, but we now have an **Edit** and **Delete** button, which do nothing. We'll fix this next:

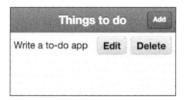

List with buttons

Making the buttons functional

In `todo-app/app/controller/Main.js`, let's edit our `control` hash to detect when one of these buttons is clicked:

```
'todo-list button[action=edit]': {
  tap: 'editTodoItem'
},
'todo-list button[action=delete]': {
  tap: 'deleteTodoItem'
},
```

Let's create the handler functions, editTodoItem and deleteTodoItem now:

```
editTodoItem: function(button, e, eOpts) {
  var store = this.getListDataView().getStore(),
    editPanel = this.getEditPanel(),
    editForm = this.getEditForm();

  editForm.setRecord(store.findRecord('id', button.getData()));

  this.showView(editPanel);
},
deleteTodoItem: function(button, e, eOpts) {
  var dataview = this.getListDataView(),
```

```
        store = dataview.getStore(),
        record = store.findRecord('id', button.getData()).erase();

    record.erase();
    store.load();
    dataview.refresh();
},
```

As with the `newTodoItem` function, these functions get the store, retrieve the record, and either display it or make changes. In the case of delete, we stay on the `List` view. In the case of edit, we load the `Edit` view.

Add two additional selectors to the refs array so that we can retrieve the `dataview` and the edit form with `getListDataView()` and `getEditForm()`, respectively:

```
listDataView: 'todo-list dataview',
editForm: 'todo-edit formpanel'
```

Finally, when the changes are made via the edit form, we need to persist these changes when the user clicks `Save`. Let's add a new function to `todo-app/app/controller/Main.js` named `saveTodoItem`:

```
saveTodoItem: function(button, e, eOpts) {
  var store = Ext.create('TodoApp.store.Item'),
    values = this.getEditForm().getValues(),
    record = store.findRecord('id', values.id);

  record.setData(values);
  record.setDirty(); // Needed otherwise update record will not sync
  store.sync();

  this.showListView();
},
```

Hook this up to the `save` button by editing its control handler:

```
'todo-edit button[action=save]': {
  tap: 'saveTodoItem'
},
```

The `Edit` view needs a couple of small changes in order to work correctly. Remember that each to-do item has a unique ID, which is generated automatically by the framework. We need to add a `hiddenfield` to the edit form so that when we submit the changes, we're able to associate the changes with an existing record.

Edit `todo-app/app/view/Edit.js` and add the following field to the form:

```
{
  xtype: 'hiddenfield',
  name: 'id'
},
```

Add a reference to the `hiddenfield` type in the requires array as a hint to Sencha Touch:

```
requires: [
  'Ext.field.Hidden'
]
```

Remove the fake value attribute from the description field in `Edit.js`. This field will be automatically populated by the store. Refresh the app and try editing and deleting your to-do items. This should work.

You may notice that we've neglected the `image` and `location` attributes in each to-do item. This was done on purpose. As we continue through the book, we'll see how to implement these features in a way that respects the offline integrity of the app that we have thus far.

 Commit your changes to Git.

Deploying the app to a phone

The final step in making our offline app real is to deploy it to a phone. You do not have to complete this section but it is a great way to see how your app will work on a phone.

We'll use iOS in this example. Note that you will need an Apple iOS Developer account ($99/year) to do this. The full instructions are available at `docs.sencha.com/touch/2.4/packaging/native_ios.html`, but we'll assume that you're running OS X and give brief instructions.

Generate an iOS certificate with the following steps:

1. Open **Keychain Access**.

2. In the application menu, navigate to **Keychain access | Certificate Assistant | Request a Certificate from a Certificate Authority**.

3. Type your e-mail and name, then choose **Save to disk** from the **Request is group** field.

4. Leave the **CA Email** field blank.

5. Click on **Continue**.

6. Save the file to your desktop.

7. Open `developer.apple.com` in your browser.

8. Log in to your account and open the **Certificates, Identifiers & Profiles | Certificates** section.

9. Click the + icon and select the **iOS App Development** certificate type.

10. Click on **Continue** and **Continue** again.

11. Click on **Choose File...** and upload the CSR that you saved earlier.

12. Click on **Generate** and wait for the iOS certificate to be generated.

13. Click on **Download**. You will need this file in a bit.

Import the certificate that you just created and configure your **App Identifier** using the Apple Developer Portal:

1. In the `Keychain Access` utility, click on **File | Import Items...**.

2. Open the `ios_development.cer` file that you just downloaded.

3. Click the **Keys** category.

4. You will notice at least one Apple ID private key. Select it. If there is more than one, choose the one that has not been revoked.

5. Click **File | Export Items...** and save the key in a `P12` format (for example, `ios_development.p12`).

6. Go back to the Apple Developer Portal and navigate to **Certificates, Identifiers & Profiles | Identifiers**.

7. Click the + icon.

8. Enter `Offline todo app` in the **Name** field.

9. In the **Bundle ID** field, enter the same reverse-domain name that you used earlier (for example, `com.example.todo-app`).

10. Click on **Continue**.

11. Click on **Submit**.

12. Make a note of the **Identifier** field as you will need this later.

Register an iOS device with the Apple Developer Portal:

1. Open **iTunes** on your computer.
2. Plug in your iOS device.
3. Click its name under the devices list.
4. Go to the **Summary** tab.
5. Click **Serial Number** to change it to the UUID.
6. Click on **Copy UUID** from the **Edit** menu.
7. Go back to the Developer Portal and open the **Devices** section.
8. Click the + icon to register your device.
9. Type a descriptive name in the **Name** field.
10. Paste the UUID in the **UUID** field.
11. Click on **Continue**.
12. Ensure that the details are correct and click on **Register**.

Create a provisioning profile using the App ID, certificate, and device that you registered earlier:

1. In the Developer Portal, navigate to the **Provisioning Profiles | Development** section.
2. Click the + icon.
3. Choose the **iOS App Development** provisioning profile.
4. Click on **Continue**.
5. Choose the App ID you created earlier and click on **Continue**.
6. Choose the certificate you created earlier and click on **Continue**.
7. Choose the device you registered earlier and click on **Continue**.
8. Type todo-app in the **Profile Name** field.
9. Click on **Generate**.
10. Click on **Download**.

Use Cordova to build and run your app in the iOS simulator:

1. Install Cordova:
    ```
    $ sudo npm install -g cordova
    ```
2. Initialize cordova in your environment:
    ```
    $ sencha cordova init com.danielsauble.TodoApp TodoApp
    ```

3. Edit `todo-app/app.json`.

4. Uncomment the platforms line and remove `android` from the string.

5. **Save** the file and exit.

6. Build your application:

   ```
   $ sencha app build native
   ```

7. Edit `todo-app/cordova/platforms/ios/TodoApp/Classes/AppDelegate.m` and add the following code to the end of the `didFinishLaunchingWithOptions` method (before the `return` statement) to prevent the app from overlapping the iOS status bar:

   ```
   if ([[[UIDevice currentDevice] systemVersion] floatValue]
     >= 7) {
     [application
       setStatusBarStyle:UIStatusBarStyleLightContent];
     self.window.clipsToBounds = YES;
     self.window.frame = CGRectMake(
       0,20,self.window.frame.size.width,
       self.window.frame.size.height-20);
   .}
   ```

8. **Save** the file and exit.

9. Build your application:

   ```
   $ sencha app build native
   ```

10. Run the application:

    ```
    $ sencha app run native
    ```

11. This will start your app in the iOS simulator.

Use Xcode to run your app on the actual iOS device:

1. To run the app on your registered hardware device, open `todo-app/cordova/platforms/ios/TodoApp.xcodeproj` in Xcode.

2. Navigate to the **Build Settings** page for your application.

3. Toggle the **All** option and scroll down to view the **Code Signing** section.

4. Choose your identity from the **Code Signing Identity** drop-down menu.

5. Plug in your phone.

6. Once the symbol libraries have been transferred, select your phone from the **Product | Destination** menu.

7. Click on **Run**.

8. You may get an error that the provisioning profile that you created earlier was not found. If so, add your developer account to Xcode first and then click on **Run**.

9. Allow the code signing to take place.

10. Unlock your phone and run the app. It should work as expected. Congratulations, you've deployed your app to iOS!

Deployed app

 Commit your changes to Git. You now have a functional (albeit basic) offline app. Let's compare it with our design principles and see how it stacks up.

Comparing and contrasting with the design principles

Keep in mind that the to-do app is not finished. As we continue through the book, we'll continue to add functionality. However, any improvements that we make will be made with the offline behavior in mind. Let's pause and evaluate the app against our list of offline principles.

How does it compare?

For each of the ten principles, we'll assign a pass/fail grade based on how well the app adheres to that principle. Then, we'll count the number of passes to see how close we came to a 10/10 score.

Give me uninterrupted access to the content I care about.

This is a pass. As the app is offline, this principle is met. We only have to worry about this when the data is separate from the device consuming that data.

Content is mutable. Don't let my online/offline status change that.

This is a pass. Again, this principle is met as there is no online functionality. You can fully edit your to-do items offline.

Error messages should not leave me guessing or unnecessarily worried.

This is a pass. Right now, the app has no error messages so this principle is met as well.

Don't let me start something I can't finish.

This is a pass. Also achieved. Without an online component, there's no need to worry about being separated from your data and being unable to complete something.

An app should never contradict itself. If a conflict exists, be honest about it.

This will be a pass. As we begin to see, when an app only has an offline functionality, it cannot violate this principle. This means that choosing offline as a starting point is a very robust decision.

When the laws of physics prevail, choose breadth over depth when caching.

This is a pass. We're offline and so we have no caching. Everything that we've created is in our local database and we don't have to share this with anybody.

Empty states should tell me what to do next in a delightful way.

This is a fail. We don't comply with this principle 100%. This could be improved. When the app first loads, what should I do first?

Don't make me remember what I was doing last. Remember for me.

This is a fail. We don't comply fully with this principle. When I close and reopen the app, it dumps me out to the to-do list, so this interaction could be improved as well. For example, if I start to create a new to-do item and my app crashes, I'd like to be brought back to that screen when I restart the app. This doesn't currently happen.

Degrade gracefully as the network degrades. Don't be jarring.

This is a pass. As the data resides on the device that consumes that data, there is no network to worry about.

Never purge the cache unless I demand it.

This is a pass. We could provide an option to reset all the data associated with the app but as you can just delete the app to accomplish the same thing, this is a pass. Once our data is separated from the app, an explicit purge option will become necessary.

Where does it need to improve?

Our first version of the app complies with 8/10 of these principles. Not bad. This evaluation should be performed early and often during development. Once a baseline is established, it's much easier to spot the regressions and areas to improve. Scoring 8/10 means we've set a strong foundation for future development.

The two items that we missed, empty states and remembering context, were missed in the design phase. As mentioned in *Chapter 1, The Pain of Being Offline*, designers often miss details such as these when the product exists as an idea in their head, a sketch on a napkin, or a high-fidelity mock-up in Photoshop. Most of the time, it takes actual development to suss out the cracks.

Fortunately, the earlier in development you spot these problems, the easier they are to fix. We'll address these in the next chapter and see if we can get a 10/10 evaluation.

Using the app as a baseline for the future

This isn't a one-time evaluation. At the conclusion of each chapter, we'll run through the principles briefly to see whether we've had any regressions. If so, we'll decide whether to fix them, and if not, explain our justification for not fixing them. In general, we expect the app to slowly improve over time as we continue to pay attention to the principles.

Our goal is to achieve perfect compliance with the principles by the end of the book. As the complexity builds, we'll be able to spot the warning signs early. In a way, these principles will serve as a meter for technical debt. We can choose not to fix the bugs that violate these principles but at the risk of more work later on.

Summary

In this chapter, we set up our development environment, designed the offline experience of our to-do app, built this experience with Sencha Touch, and deployed it to our device. We also evaluated the offline experience of the app against our principles and found areas where the experience could be improved.

In the next chapter, we'll continue to add functionality, improve some of the broken or neglected areas of the app, design the online flow of the app, and walk through this flow in order to see well how it transitions between the offline and online states. Then, we'll use Cloudant as an online store for our to-do data.

3
Designing Online Behavior

Now that we have built an offline to-do app, let's go one step further and decide what the full online experience will be. As we progress, our goal is to migrate the app to the Internet without compromising its offline behavior.

It's useful to think about this in terms of a bunker and beach house analogy. An offline-only app is like a hardened bunker: invulnerable to what's happening above the ground but quite isolated. It's not a place you'd want to spend all your time, though you'll be glad to be able to retreat there when necessary.

An online-only app, in contrast, is more like a beach house. It's bright and cheery; the kind of place you'd want to invite your friends to hang out. However, if an earthquake or tsunami should happen, you'll quickly forget about all the nice aspects of your building as you flee for the hills.

What we're trying to do is marry these two concepts and create a hybrid building. A beach house when things are great and a bunker when you need to hunker down. The problem is that most developers optimistically focus on the bright, cheery outdoors and then find it difficult to retrofit their foundations to support a hardened *bunker mode*.

Our approach is the opposite. In *Chapter 2*, *Building a To-do App*, we built the beginnings of our *bunker mode*. Now, we can start erecting our above-ground abode. We won't neglect the bunker, however. It's not finished. Now that the essentials are in place, we can easily go back and forth and make modifications to either building as needed.

This continuous iteration is one of the most important aspects of developing a robust experience. As you build the beach house, you'll discover aspects that you want available whether above- or below-ground. When this happens, we'll modify both the offline underpinnings and online experience in tandem.

Designing the online workflow

As you'll recall, this is the offline workflow that we designed in *Chapter 2, Building a To-do App*:

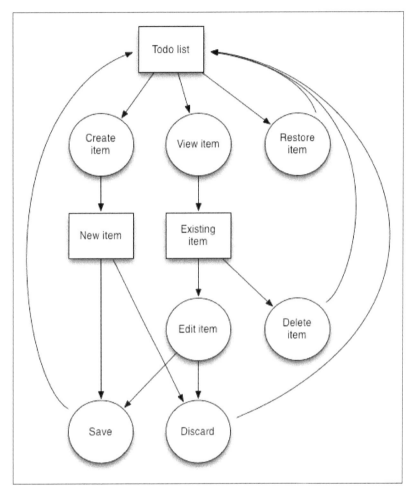

Workflow

This workflow is the offline nucleus of our app. We may add additional functionalities but it is not allowed to impinge on the offline integrity of this basic workflow. We will visualize this nucleus as follows in the next few diagrams, with the offline functionalities shaded in grey:

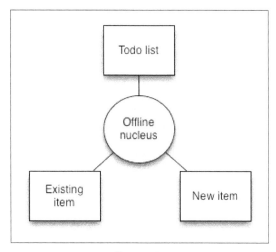

Offline nucleus

The existing screens are available for us to extend in the online mode but the core workflow remains unchanged. What additional functionalities should we enable while online?

Let's review the user stories from *Chapter 1, The Pain of Being Offline*:

1. Items should display while offline.
2. Keep all my devices in sync.
3. Don't keep me waiting.
4. Sharing with friends should be bulletproof.
5. Safeguard my data.

Of these user stories, 2, 4, and 5 have requirements that are not currently satisfied by our offline-only experience. Let's see how we can extend the current workflow to accommodate these stories.

Keep all my devices in sync

Susan owns multiple devices. She would like to install the to-do app on each device. These devices should be synchronized with each other so that her to-do list is available, no matter which device she chooses. She wants one list, not several.

Synchronization is largely a background operation. We shouldn't expect Susan to click a **Sync** button whenever she adds a to-do item. This operation should happen automatically. However, if Susan wants to be sure that her items are synced, we should provide an affordance to do so. This affordance should give her reassurance about which items have been synced so that she doesn't feel compelled to click it in normal usage:

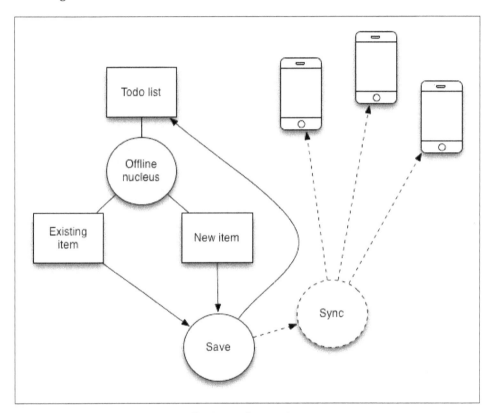

Device synchronization

In our updated workflow diagram, synchronization is triggered on **Save**. As it is an automatic operation, not normally triggered by a human, I've outlined the related elements with dashes. On **Save**, this **New item** or **Existing item** is sent to the other devices connected to the current list.

Sharing with friends should be bulletproof

Susan is an organized person. She can't force her friends to be organized as well but she can employ tools to help them be more organized. Each of her close friends has installed the to-do app on their phones. When they share lists with each other, the items on these lists are synced automatically.

Our offline app has only one to-do list. We'll need to provide a way to manage additional lists, share them easily with others, and keep them all in sync. As this is a similar, but larger, problem to the second user story, we'll solve this next. Also, as we can't guarantee that all of her friends have Internet connectivity, we'll need to investigate ways to synchronize their devices when Internet is not available:

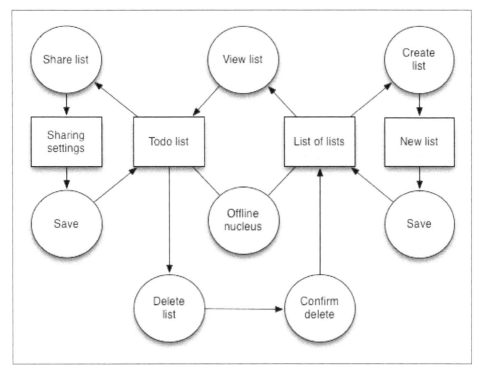

List management

I've added a couple of additional views (in grey) with their associated actions to the **Offline nucleus**: **List of lists** and **New list**. The other new view, **Sharing settings**, is only available online or with appropriate online alternatives (for example, Bluetooth, Wi-Fi, and so on).

Safeguard my data

Susan worries constantly about data loss. An offline app is pretty safe but she doesn't always remember to back up her devices. If she drops her phone in the pool, her data is gone. Of course, it is unlikely that she will drop all of her devices in the pool at the same time, so synchronization can help us somewhat. However, in this unlikely event, can we do anything to preserve her data, regardless?

Most synchronization tools have a remote server where all the data is stored. When a device needs to retrieve or update the data, they ask the server. As we can't assume that Internet is always available, we should allow offline synchronization but employ an online data store to serve as an authoritative backup.

There is an additional complexity behind all of these stories. We'll need lightweight management to differentiate between different people and to transfer their data to the server. We must design this in such a way that the app does not become inoperable when it goes offline:

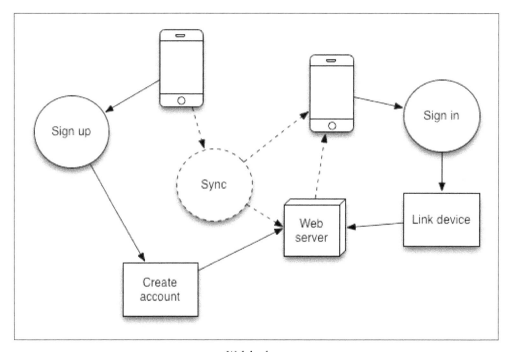

Web backup

Note that all of these views are online views. If the app is offline, none of these operations can be performed. There are two parts to this workflow. On the left, a device may **Sign up** for a new account, which allows it to sync to-do items to a **Web Server**. On the right, a device may log in to an existing account, linking it to the **Web Server** as well.

When items are synced, they will first attempt to be sent directly to all the linked devices. They will then be sent to the **Web Server**. If a linked device is offline during this process, it will download any to-do items that were synced in the interim when it comes online.

The online/offline transition

Now that we've designed the online and offline behavior, let's talk about the transition between these two states. As the ninth principle states, the UI transition as the network degrades should not be jarring. How is this reflected in our workflow diagrams?

As mentioned, an online functionality is highlighted in white, while an offline functionality is highlighted in grey. The transition points are the lines between white and grey. We need to think about what happens when someone is in the white space and their Internet connection goes down, or when they are in the grayspace and want to do something that is reserved for online use only. How do we be transparent about what's going on?

Being informative about what can or cannot be done

When people are offline, they may want to do something that may only be done online. This is a reasonable desire but we can't defy the laws of physics, so what do we do? There are a few strategies that we can take. Going from the most restrictive to least restrictive, they are as follows:

* Tell people that they have to wait until they're online
* Let people set a reminder for themselves to complete the activity when online
* Let people complete the offline parts, leaving the online parts for later

In general, allow people to do as much as possible offline. Let them save their progress and complete or submit it when online. As we continue, we'll choose one of these three strategies for each piece of functionality that we will implement.

Graceful degradation

When people are online, they may be in the middle of something that's important or critical. As we cannot guarantee that their Internet connection will remain reliable until they're done, we need strategies to cope with this. There are a few approaches that we can take:

- Warn people that the Internet connection is weak

- Anticipate an impending lack of connectivity and give people a time estimate

- Make the app less bandwidth-hungry when the bandwidth is limited

These strategies are not on a continuum. They can be employed separately or together, depending on the importance of the task at hand. We'll choose from these strategies as we continue to implement the app.

Recovering when things go horribly wrong

When people are online, sometimes the unexpected happens and the Internet connection just dies, completely, without warning. If it doesn't come back right away, we don't want to force people to keep the app open. We need to give them closure and a sense of security about their work so that they don't stress about losing it. There are a few strategies here as well:

- Tell people what happened and save a draft of what they were working on

- Let people *submit* their work and promise to post it when online

- Let people know when the Internet died so that they know exactly what was posted and what wasn't

As we'll discuss in *Chapter 6, Be Eventually Consistent*, split-brain is another problem that can cause chaos when a person does something offline. The app tries to reconcile that change when the app goes back online and fails. We'll discuss this problem separately in that chapter.

Example scenarios

Now, we'll pick a few scenarios from our workflows and figure out how best to address them with the strategies mentioned previously. We won't cover all of the scenarios but we'll reference this list later in the book when required.

Susan creates a to-do item while offline

Creating a to-do item is part of the offline nucleus. Thus, we shouldn't force Susan to click a button to acknowledge that it hasn't been synced. Instead, let's use lightweight styling to tell her what's going on and remove the styling when the app goes online and synchronizes successfully:

- After Susan creates a to-do item, the app tries to synchronize itself

- As the app is offline, this fails

- Instead, it shows the to-do item in light grey with the word *pending* underneath to indicate that it has not been synchronized

- When Susan goes online, the app synchronizes successfully and the to-do item reverts to the normal styling treatment

Susan tries to share a to-do item while offline

Unlike creating an unshared to-do item, this is not something that can be done successfully while offline. Susan has a number of shared lists. As the primary use case of these lists cannot be achieved while offline, we should give her more notice when we run in trouble but still don't require interaction on her part:

- After Susan creates a shared to-do item, the app tries to synchronize itself.

- As the app is offline, this fails.

- Instead, it shows the to-do item in the offline treatment as before. In addition, it puts a bold *1 Item Not Shared* message at the top of the list in order to make Susan aware that her friends can't see it yet.

- When Susan goes online, the app synchronizes successfully, the to-do item reverts to the normal styling treatment, and the message at the top of the list disappears.

Susan synchronizes her lists over a weak connection

Susan hates waiting. In her to-do app, she interacts most with the text, then the location, and then the photos and videos attached to each item. She wants these attributes downloaded in this order and available for use immediately. She doesn't want to wait for the least-used attribute to download before she can access the most-used attribute:

- After Susan opens the to-do app, it starts to download the updated lists.

- First, it downloads the text for each item, making this text available to be read or modified as soon as it is downloaded.

- Next, it downloads a map for each item (that has a location), making this data accessible as well.

- Finally, it downloads the photos or videos attached to each item.

- While any piece of information is downloading, it gets a spinner. This spinner is removed once the download is complete.

Susan adds a video attachment over a weak connection

We shouldn't prevent Susan from doing this but we should warn her that it might take awhile. In fact, we should give her an estimate so that she can judge this for herself.

- Susan edits an existing to-do item

- In the photo/video section, we mention that the Internet connection is weak and so adding new media could take awhile

- She clicks Add Photo/Video anyway and chooses a video to attach

- As the video uploads, we show how long the upload will take

- Susan doesn't want to wait that long so she closes the app

- When she reopens the app, we show the video upload as paused

- When the Internet connectivity is 4G, LTE, or Wi-Fi, we resume the video upload automatically

Susan adds multiple to-do items while offline

As in a single case, we shouldn't make it difficult for Susan to do this. However, if there are more than five items in a list that are not synchronized, we should let her know that she is accumulating significant unsaved work.

- Susan adds more than five to-do items while offline

- We put a bold *5+ Items Not Synchronized* message at the top of the list

- If she clicks on the message, we display *We'll save your to-do items as soon as you're online. Consider running a backup if you're going to be offline for awhile*

- When Susan runs a backup or goes online, we reset the warning message

Adding an online data store

In the next chapter, we will start to improve the to-do app to support our revised workflow. Before we do this, we need a place to store our data online, which supports device syncing and user sharing. We'll use **CouchDB** for this.

CouchDB is a **NoSQL** database that makes certain things very easy: database synchronization, data access (via a built-in REST endpoint), and user management. There are several places to get a free CouchDB database online. We'll use **IBM Cloudant** to provide this service.

Creating a Cloudant account

IBM Cloudant is an online database host. It's free as long as your monthly usage is less than $50. One alternative is **Iris Couch** (www.iriscouch.com), which works in basically the same way. For now, let's use Cloudant:

1. Open www.cloudant.com in your browser.
2. Click on **Sign Up**.
3. Choose your credentials.
4. Click on **I agree, sign me up**.
5. Click on **Add New Database**.
6. Name your database lists.
7. Click on **API URL** to copy the link to your new database.
8. Paste the link in a new tab.
9. Enter your credentials, if requested, and authenticate.
10. You should see an output similar to this:

Working CouchDB

11. Go back to your Cloudant dashboard.

12. Select your lists database, if not already selected.

13. Click **Permissions**.

14. Click **Generate API key**.

15. Save the key and password. You will use these to authenticate in the future so that you don't need to share your master password.

16. Check the **Reader**, **Writer**, and **Replicator** boxes for the key that you created.

Your new database is ready to go. Next, let's define our data structures and decide what the constraints should be.

Data structures and validation

Before we write code, let's go through the data that we need. I'll mock this out in **JSON** as it's easy to read and what we'll end up using anyway.

We need a structure to store each to-do list. The first attribute is the unique ID for the structure. The list of to-do items is an array under `items`:

```
{
  id: <int>,
  name: <string>,
  owner: <int>,
  collaborators: [<int>],
  items: [
    {
      id: <int>,
      description: <string>,
      media: <string>,
      latitude: <number>,
      longitude: <number>
    },

  ]
}
```

As CouchDB comes with authentication built-in, we don't need to create a separate structure for the user information. Pretty sweet!

That's it! A to-do list has references to the people that own and collaborate around it and a reference to the list that owns it. In a traditional app, we would let the database manage these relationships and any constraints, but it's simpler to keep this logic in the app. Obviously, if you wanted to build a production-ready application, you would want to enforce all this on the server side; so keep this in mind.

Our web service should allow people to perform basic CRUD operations on these structures. This data has the following constraints:

- A to-do list must have a name that is a string.
- A to-do list must have an owner.
- The owner field must be set to the current user's ID.
- The owner field cannot be changed once it is set.
- Only the owner can delete the list.
- The collaborators field, if present, must be an array of strings.
- Only owners and collaborators can view and edit a to-do list.
- The items field, if present, must be an array of objects.
- Each to-do item must have a description that is a string.
- Each to-do item must have an ID that is a string.
- Each to-do item can have a media field. If present, it must be a string.
- Each to-do item can have a latitude and longitude field. If one is present, the other must be as well and both must be numbers.

Now that we've outlined the requirements, let's write some code.

Writing a validation function

In the Cloudant dashboard, perform the following steps:

1. Select your database.
2. Click the + icon next to **All Design Docs**.
3. Click **New Doc** to create a new design document.
4. Prefix the value of the `_id` key with `_design/`. (This should appear as `_design/e106bcd....`)

You've created a new design document. In CouchDB, we can create something called a `validate_doc_update` function, which validates any requests made to the database. This is used to ensure that any validation requirements are met.

To do this, add another key to the design document named **validate_doc_update**. Then, click the edit button next to the field to input the validation function:

Cloudant field editor

In the editor that appears, create an empty function as follows:

```
function(newDoc, oldDoc, userCtx) {
}
```

If you saved this as is, any request made to the database would succeed, which is the default behavior. Now, insert each of the following sections into the function in the order given:

```
if (!newDoc['name']) {
  throw({forbidden: 'todo list must have a name'});
}
if (typeof newDoc['name'] !== 'string') {
  throw({forbidden: 'todo list name must be a string'});
}
```

This ensures that each to-do list has a name attribute and it is set to a string (not an array, a number, or an object).

Now, let's set requirements for ownership of the list:

```
if (!newDoc['owner']) {
  throw({forbidden: 'must have an owner field'});
}
if (!oldDoc && newDoc['owner'] != userCtx.name) {
  throw({forbidden: 'must set the owner field to your own ID'});
}
if (oldDoc && newDoc['owner'] != oldDoc['owner']) {
  throw({forbidden: 'may not change the owner field once set'});
}
if (newDoc['_deleted'] && newDoc['owner'] != userCtx.name) {
  throw({forbidden: 'only the owner may delete the list'});
}
```

This ensures that every list has an owner, which the owner cannot change, the person who creates the list is the owner, and only the owner has the privilege to delete the list. This is very restrictive and we may want to loosen these requirements in the future. For example, it might be useful to transfer ownership of a list to another person or let a person create lists on behalf of someone else.

Next, let's set some requirements for collaboration:

```
if (newDoc['collaborators']) {
  if (Object.prototype.toString.call(newDoc['collaborators']) !==
    '[object Array]') {
    throw({forbidden: 'collaborators field must be an array'});
  }
  for (var i = 0; i < newDoc['collaborators'].length; ++i) {
    if (typeof newDoc['collaborators'][i] !== 'string') {
      throw({forbidden: 'collaborators field must be an array of
strings'});
    }
  }
}
```

This ensures that the `collaborators` field, if set, is an array that contains strings (the IDs of the other users with permission to view and edit this list). Now, let's ensure that only the `owner` and `collaborators` can view and edit lists:

```
if (newDoc['owner'] != userCtx.name && newDoc['collaborators'] &&
  newDoc['collaborators'].indexOf(userCtx.name) === -1) {
  throw({forbidden: 'only the owner and collaborators may view or
edit'});
}
```

Pretty easy! Next, let's define the fields in each to-do item:

```
if (newDoc['items']) {
  if (Object.prototype.toString.call(newDoc['items']) !==
    '[object Array]') {
    throw({forbidden: 'items field must be an array'});
  }

  for (var i = 0; i < newDoc['items'].length; ++i) {
    if (typeof newDoc['items'][i] !== 'object') {
      throw({forbidden: 'items field must be an array of
        objects'});
    }
    if (!newDoc['items'][i]['id']) {
      throw({forbidden: 'todo item must have an id field'});
```

```
  }
  if (typeof newDoc['items'][i]['id'] !== 'string') {
    throw({forbidden: 'todo item id must be a string'});
  }
  if (!newDoc['items'][i]['description']) {
    throw({forbidden: 'todo item must have a description
      field'});
  }
  if (typeof newDoc['items'][i]['description'] !== 'string') {
    throw({forbidden: 'todo item description must be a string'});
  }
if (newDoc['items'][i]['media'] && typeof newDoc['items'][i]['media']
!== 'string') {
    throw({forbidden: 'todo item media field must be a string'});
  }
```

Each to-do item must have an id field and a description field, both strings. They may optionally have a media field, also a string.

The location is a bit more complicated. Let's add this next:

```
if (newDoc['items'][i]['latitude'] &&
  !newDoc['items'][i]['longitude']) {
  throw({forbidden: 'todo item is missing longitude'});
}
if (newDoc['items'][i]['longitude'] &&
  !newDoc['items'][i]['latitude']) {
  throw({forbidden: 'todo item is missing latitude'});
}
if (newDoc['items'][i]['latitude'] && typeof
  newDoc['items'][i]['latitude'] !== 'number') {
  throw({forbidden: 'todo item latitude must be a number'});
}
if (newDoc['items'][i]['longitude'] && typeof
  newDoc['items'][i]['longitude'] !== 'number') {
  throw({forbidden: 'todo item longitude must be a number'});
}
}
}
```

A to-do item may optionally have a location defined by the latitude and longitude fields. It doesn't make sense to have only one of these, so we require both if either is present. They must also be numbers.

That's it for the requirements. Pretty easy, huh? Click **Save** and **Save** again and your validation function is now live. Don't worry about how you'll manage user accounts for now. In the next chapter, we'll use a plugin to add user management to our app, such as creating accounts, logging in, changing passwords and e-mail addresses, and deleting accounts.

Testing the validation function

Now, let's check our database to see how well it handles validation. To do this, you'll need your command-line terminal, curl, and API key and password that you generated earlier. Type the following commands to save the redundant bits of the curl commands for future use:

```
$ export DOC=https://[key]:[password]@username.cloudant.com/lists/test
$ export ARGS="-X PUT -H 'Content-Type: application/json'"
```

Now, let's test our validation function. We'll start by passing a blank object and then address the errors individually until they pass:

```
$ curl $ARGS -d '{}' $DOC
{"error":"forbidden","reason":"todo list must have a name"}
```

As expected, this fails. Let's add a name:

```
$ curl $ARGS -d '{"name": {}}' $DOC
{"error":"forbidden","reason":"todo list name must be a string"}
```

Let's change the name to a string:

```
$ curl $ARGS -d '{"name": "household"}' $DOC
{"error":"forbidden","reason":"must have an owner field"}
```

The name validates; now for the owner field:

```
$ curl $ARGS -d '{"name": "household", "owner": {}}' $DOC
{"error":"forbidden","reason":"must set the owner field to your own ID"}
```

The owner can't be just anybody, so let's change the ID to our API key:

```
$ curl $ARGS -d '{"name": "household", "owner":
"thistrompecoughtsideneet"}' $DOC{"ok":true,"id":"test","rev":"1-07b69f3e
296bb5bec26670e002bbbe0a"}
```

This worked. So, the minimum fields necessary to create a new to-do list are the `name` and `owner`. Now, let's verify that the `owner` field is immutable:

```
$ curl $ARGS -d '{"name": "household", "owner": "differentowner", "_rev":
"1-07b69f3e296bb5bec26670e002bbbe0a"}' $DOC

{"error":"forbidden","reason":"may not change the owner field once set"}
```

You can't, as designed. Now for the `collaborators` field; it's optional, but once we decide to set it, what are the restrictions?

```
$ curl $ARGS -d '{"name": "household", "owner":
"thistrompecoughtsideneet", "collaborators": {}, "_rev": "1-07b69f3e296b
b5bec26670e002bbbe0a"}' $DOC{"error":"forbidden","reason":"collaborators
field must be an array"}
```

The `collaborators` field only takes an array. Let's give it one:

```
$ curl $ARGS -d '{"name": "household", "owner":
"thistrompecoughtsideneet", "collaborators": [{}], "_rev": "1-07b69f3e296
bb5bec26670e002bbbe0a"}' $DOC

{"error":"forbidden","reason":"collaborators field must be an array of
strings"}
```

Each element of the `collaborators` array must be a string. Let's provide this:

```
$ curl $ARGS -d '{"name": "household", "owner":
"thistrompecoughtsideneet", "collaborators": ["george"], "_rev": "1-07b69
f3e296bb5bec26670e002bbbe0a"}' $DOC

{"ok":true,"id":"test","rev":"2-e70405607c52024ca615da1a8ed9f0b1"}
```

This worked. Perfect! Up until now, all of our edits have been with the owner of the list. Let's make sure that the collaborators can edit the lists as well (but nobody else). To do this, let's create a couple of additional users in the Cloudant dashboard:

1. Select the **lists** database.
2. Click **Permissions**.
3. Create two new API keys. (Click **Generate API Key** twice.)
4. For each key, save the key and password for future reference.
5. Make sure that both the users have read and write permissions.

Now, add the first user to the to-do list as a collaborator:

```
$ curl $ARGS -d '{"name": "household", "owner":
"thistrompecoughtsideneet", "collaborators": ["george"], "_rev":
"5-e4e63ef94ea8e1dceb7d47d5ff0feba3"}' $DOC

{"ok":true,"id":"test","rev":"6-1a4b906bfac10b11303e5ddaebbd9cb6"}
```

If we set the validation function correctly, the first user should be able to edit the list but not the second user:

```
$ export DOC=george:password@djsauble.cloudant.com/lists/test

$ curl $ARGS -d '{"name": "household", "owner":
"thistrompecoughtsideneet", "collaborators":
["mrsedgerindogedgedgerady"], "_rev":
"7-ecb03521c54567d78ec262070d06f6f7"}' $DOC

{"ok":true,"id":"test","rev":"8-20d3d307d5d6f93d322eb999faa7249b"}
```

As expected, the first user is able to retrieve the document. How about the second user?

```
$ export DOC=clarissa:password@djsauble.cloudant.com/lists/test

$ curl $ARGS -d '{"name": "household", "owner":
"thistrompecoughtsideneet", "collaborators":
["mrsedgerindogedgedgerady"], "_rev":
"8-20d3d307d5d6f93d322eb999faa7249b"}' $DOC

{"error":"forbidden","reason":"only the owner and collaborators may view
or edit"}
```

The second user cannot do so. This is exactly what we want. All the to-do items are public, in that anyone can view them, but only the owners or collaborators can make changes.

Finally, let's validate that only the owner can delete a to-do list:

```
$ export DOC=mrsedgerindogedgedgerady:SNoJan2kuPSvVnlw7LeoBgSF@djsauble.
cloudant.com/lists/test

$ curl $ARGS -d '{"_deleted": true, "name":
"household", "owner": "thistrompecoughtsideneet",
"collaborators": ["mrsedgerindogedgedgerady"], "_rev":
"8-20d3d307d5d6f93d322eb999faa7249b"}' $DOC

{"error":"forbidden","reason":"only the owner may delete the list"}
```

Collaborators cannot delete a list that they don't own. That's good.

```
$ export DOC=https://thistrompecoughtsideneet:VW6uPL3kNHbdlHuoovjWVCUX@
djsauble.cloudant.com/lists/test

$ curl $ARGS -d '{"_deleted": true, "name":
"household", "owner": "thistrompecoughtsideneet",
"collaborators": ["mrsedgerindogedgedgerady"], "_rev":
"8-20d3d307d5d6f93d322eb999faa7249b"}' $DOC

{"ok":true,"id":"test","rev":"9-ca30bb170b17ad818acf90b6f447c3bf"}
```

The owners can delete their lists. Great!

We're done! The validation function might seem like overkill but it serves a very important purpose. Yes, you could implement client-side validation in your app and call it good. However, as CouchDB exposes a REST API that anyone can use, it's a good idea to put the validation as close to the data as possible. It assures us that regardless of where the data comes from, it's well-formed and easily consumed by any app we write.

As you went through the commands, you may have noticed an extra attribute named _rev. The revision attribute is used to track the changes. Similar to Git, all the older versions of a document are preserved unless you explicitly delete them. When we make updates to a document, we must specify which revision is the latest. This attribute changes every time a successful update occurs.

The reason for the attribute is to help us solve the problem that occurs when two separate databases, with different changes for the same document version, try to reconcile their changes. We'll walk through this problem (and the solution) in *Chapter 6, Be Eventually Consistent*.

Implementing image support

We left a few features unfinished in the last chapter. Currently, you cannot attach images to a to-do item or choose a location from the map. Let's implement the first of these now.

Installing the Cordova camera plugin

As Sencha Touch is a platform-agnostic framework, it doesn't have hooks in any of the native capabilities of your devices, including the camera. The downside is that there isn't a platform-independent way to attach photos. You have to implement separate solutions for different platforms. To keep things simple, we'll implement a solution for iOS. You'll be able to continue using the app with your desktop browser; you just won't be able to attach photos.

In *Chapter 2, Building a To-do App* we used Cordova to package our Sencha Touch app and deploy it to iOS. Cordova provides you with the ability to integrate with your phone's native capabilities through a plugin system. We'll use the camera plugin to give the app access to the images on your device.

Open your command-line terminal. In the `todo-app/` project folder, switch to the `cordova/` directory and install the plugin:

```
$ cordova plugin add cordova-plugin-camera
```

This will install the necessary files in your project folder. Now, we can access the methods that it exposes directly in our app. Let's open Sublime Text and make the necessary changes.

Creating a custom image component

In `new.js` and `edit.js`, we added a placeholder for the image functionality wrapped in a fieldset. We'll remove these placeholders and create a new file containing a custom image component along with the logic to select and save images. Create a new file under `todo-app/app/views/` named `image.js` and make the following changes.

1. First, provide the name of the new component and derive it from `Ext.form.FieldSet`:

```
Ext.define('TodoApp.view.Image', {
  extend: 'Ext.form.FieldSet',
  alias: 'widget.todo-image',
});
```

2. Under the `alias` attribute, define the configuration for the component, including the title of the fieldset and components that live under the fieldset:

```
config: {
  title: 'Image',
  items: [
    {
      xtype: 'hiddenfield',
      name: 'media'
    },
    {
      xtype: 'panel',
      layout: 'fit',
      html: 'No image loaded'
    },
    {
      xtype: 'button',
      text: 'Select'
    },
    {
      xtype: 'button',
      text: 'Remove',
      hidden: true
    }
  ]
},
```

The `hiddenfield` component will hold the Base64-encoded image, to be submitted as a part of the form. The other controls are used to display and select this image. The buttons, select and reset, are mutually exclusive; only one will be shown at a time.

3. Under the `config` block, let's add functions to select and remove images. We'll link to these from the button handlers in a bit. First, remove the following:

```
removeImage: function(scope) {
  scope.down('hiddenfield').setValue('');
  scope.down('panel').setHtml('No image loaded');
  scope.down('button[text=Select]').setHidden(false);
  scope.down('button[text=Remove]').setHidden(true);
},
```

4. This function clears the `hiddenfield` component, provides text to indicate that no image is loaded, and shows the `Select` button while hiding the `Remove` button. Let's add the `selectImage` function next:

```
selectImage: function(scope) {
  navigator.camera.getPicture(
    function(dataUrl) { // Success
      var media = 'data:image/jpeg;base64,' + dataUrl;
      scope.down('hiddenfield').setValue(media)
      scope.down('panel').setHtml('<img src="' + media + '"
        alt="todo image" width="100%"/>');
      scope.down('button[text=Select]').setHidden(true);
      scope.down('button[text=Remove]').setHidden(false);
    },
    function(message) { // Failure
      scope.down('panel').setHtml(message);
    },
    { // Options
      quality: 50,
      destinationType: navigator.camera.DestinationType.DATA_URL,
      sourceType: navigator.camera.PictureSourceType.PHOTOLIBRARY,
      CameraUsesGeolocation: true
    }
  );
},
```

This function is a little more complicated. It uses the Cordova camera plugin that we installed earlier to select an image from your device. If this is successful, it sets the value of `hiddenfield` to the Base64-encoded image, displays the image in an `img` tag, and shows the `Remove` button. It also specifies a few options to decrease the image size, request an image from the photo library, and use the geolocation information from the JPEG, if available.

5. The last thing that we need to do in this file is hook up the button handlers with these two functions. In the `config` block, add `handler` to the `Select` button:

    ```
    {
      xtype: 'button',
      text: 'Select',
      handler: function(button) {
        var parent = button.up('todo-image');
        parent.selectImage(parent);
      }
    },
    ```

6. This causes the `selectImage` function to be called whenever the `Select` button is clicked, with a reference to the parent (`Ext.form.FieldSet`). Now, do the same for the `Remove` button configuration:

    ```
    {
      xtype: 'button',
      text: 'Remove',
      hidden: true,
      handler: function(button) {
        var parent = button.up('todo-image');
        parent.removeImage(parent);
      }
    }
    }
    ```

Great, you've defined a custom image control. Now, let's wire it up to the rest of the to-do app.

Enabling image support

As mentioned, `new.js` and `edit.js` contain placeholders for the image controls. Remove the image fieldset block in these files and replace it with the following code:

```
{
  xtype: 'todo-image'
},
```

Then, to load the necessary classes, add a reference to the image component in the `requires` array for each file:

```
requires: [
   'TodoApp.view.Image'
],
```

That's it! Now, when you run your app, you can **Create** a new item, select an image from your device, and attach it to the item:

To-do app with image support

There's just one thing left to do. When you edit the item, the image doesn't display. The image is automatically loaded in the `hiddenfield` component, but we haven't provided the logic to automatically display the image. Let's do this now. Edit `todo-app/app/controller/main.js` and modify the `editTodoItem` function:

```
editTodoItem: function(button, e, eOpts) {
  var store = this.getListDataView().getStore(),
  editPanel = this.getEditPanel(),
  editForm = this.getEditForm(),
  imagePanel = editForm.down('todo-image').down('panel'),
  record = store.findRecord('id', button.getData()),
  mediaData = record.get('media');

  editForm.setRecord(record);

  // Show the associated image
  if (mediaData) {
    imagePanel.setHtml('<img src="' + record.get('media') + '"
      alt="todo image" width="100%"/>');
  }

  this.showView(editPanel);
},
```

Now, run your app and edit an item with an image attached. Notice how the image is displayed correctly.

Implementing mapping support

The map is the other big chunk of functionality that's currently missing. When you create a new to-do item, you should have the option to select a point on the map. When you near this point, your phone will remind you of the related item on your to-do list.

Creating a custom map component

As with the image selector, we'll create a custom map component to replace the temporary fieldset that we created when we were mocking the views. Create a new file in `todo-app/app/view/` named `Map.js` and give it the basic component structure:

```
Ext.define('TodoApp.view.Map', {
  extend: 'Ext.form.FieldSet',
  alias: 'widget.todo-map',
  requires: [
    'Ext.Map'
  ]
});
```

Sencha Touch provides a map component, which wraps the Google map API. We require the `Ext.Map` class here but we also need to include the JavaScript that is backing this API. Add the following element to the `js` array in `todo-app/app.json`:

```
{
  "path": "http://maps.google.com/maps/api/js?sensor=true"
}
```

Now, back in `Map.js`, add a `config` attribute to define basic `config` options:

```
config: {
  title: 'Location',
  items: []
}
```

In the items array, add fields to hold the values for `latitude` and `longitude`:

```
{
  xtype: 'hiddenfield',
  name: 'latitude'
},
{
  xtype: 'hiddenfield',
  name: 'longitude'
},
```

Now, add a placeholder `panel` to hold the `Ext.Map` instance. We'll add the map dynamically, for reasons that we'll discuss later:

```
{
  xtype: 'panel',
  layout: 'fit',
```

```
      width: '100%',
      height: 300
    },
```

Finally, add buttons to `Set` and `Clear` the location for a given to-do item:

```
    {
      xtype: 'button',
      text: 'Set'
    },
    {
      xtype: 'button',
      text: 'Clear',
      hidden: true
    }
```

Open `todo-app/app/view/Edit.js` and replace our placeholder fieldset with a reference to the new map component:

```
{
  xtype: 'todo-map'
}
```

Add a dependency to the `requires` array at the top of the file:

```
    'TodoApp.view.Map'
```

Now do the same for `todo-app/app/view/New.js`. Pretty simple!

Adding logic to the view

Now that we've defined the components in our custom view, let's add the logic necessary to make the map load and display correctly. First, we need to add a way to set a marker on the map indicating the alert location. Add the following function:

```
    setMarker: function(me, latitude, longitude) {
      var map = me.down('map').getMap(),
        position;

      if (map.mapMarker) {
        map.mapMarker.setMap(null);
      }

      if (latitude && longitude) {
        position = new google.maps.LatLng(latitude, longitude);
```

```
    } else {
      position = map.getCenter();
    }

    map.mapMarker = new google.maps.Marker({
      position: position,
      map: map
    });

    me.down('button[text=Set]').setHidden(true);
    me.down('button[text=Clear]').setHidden(false);
    me.down('map').setUseCurrentLocation(false);
    me.down('map').setMapOptions({
      center: map.mapMarker.getPosition(),
      draggable: false
    });
    map.mapCenter.setVisible(false);
  },
```

There's a lot going on here but it's pretty simple. First, we reset any previous marker on the map. Next, if a `latitude` and `longitude` is provided, we create the marker at that position; otherwise, we create it at the center of the map. Next, we instantiate the actual marker, pointing it at the position and the map itself. Finally, we the show/hide buttons, fix the map to its current position, and hide the marker placeholder (more on this later).

Next, we need a method to clear the marker from the map and use a placeholder instead (to show where the marker would be positioned, if set):

```
  clearMarker: function(me) {
    var map = me.down('map').getMap();

    if (map.mapMarker) {
      map.mapMarker.setMap(null);
    }

    me.down('button[text=Set]').setHidden(false);
    me.down('button[text=Clear]').setHidden(true);
    me.down('map').setMapOptions({draggable: true});
    map.mapCenter.setVisible(true);
  }
```

This method also clears any existing marker. It frees the map so that it can be dragged around. It also displays the placeholder for the center of the map. Next, we'll implement the method that sets this up.

So, we have methods that set and clear our map marker. Now, let's implement a method that initializes the map when it first renders:

```
onMapRender: function(obj, map) {
  var mapPanel = obj.up('todo-map');

  if (!map.mapCenter) {
    map.mapCenter = new google.maps.Marker({
      map: map,
      opacity: 0.5
    });
    map.mapCenter.bindTo('position', map, 'center');
  }

  if (!mapPanel.mapRendered) {
    map.mapRendered = true;
    mapPanel.onMapAdd(obj, map);
  }
},
```

When someone sets a location on a to-do item, a marker appears at the center of the map. However, we want people to know precisely where the marker will appear. To solve this, we add a ghost marker at the center of the map. As the map is dragged about, the marker updates its position. Once the location has been set, the ghost marker is hidden and the real marker is displayed in its place.

The other purpose of this method is to kick-start the onMapAdd method, which we will add next. This method checks for the latitude and longitude data in the form and sets the marker, if it exists. Let's implement this method now:

```
onMapAdd: function(obj, map) {
  var mapPanel = obj.up('todo-map'),
    longitude =
      mapPanel.down('hiddenfield[name=longitude]').getValue(),
    latitude =
      mapPanel.down('hiddenfield[name=latitude]').getValue();

  if (map.mapRendered) {
    if (longitude && latitude) {
      mapPanel.setMarker(mapPanel, latitude, longitude);
    } else {
      mapPanel.down('map').setUseCurrentLocation(true);
      mapPanel.clearMarker(mapPanel);
    }
  }
},
```

As mentioned, this method checks for `latitude` and `longitude` in the form. If the map has been rendered, we set the marker appropriately. If no data has been provided, we let the map snap to the user's current location in order to make it easier to select a location in the future.

If the map has not been rendered, we do nothing. This is fine as the method will be called as soon as the map has rendered. (See `onMapRender`.)

Now that we've defined these methods, let's hook them up to the event handlers that call them. In the items array of `Map.js`, add event handlers to the panel and two buttons:

```
listeners: {
  add: function(obj, item, index) {
    var parent = obj.up('todo-map');
    parent.onMapAdd(obj, item.getMap());
  }
}
```

When a map is dynamically added to the panel, display it and set the markers appropriately. Now, for the `Set` button, do the following:

```
{
  xtype: 'button',
  text: 'Set',
  handler: function(button) {
    var parent = button.up('todo-map'),
    map = parent.down('map').getMap();

    parent.setMarker(parent);

    var position = map.mapMarker.getPosition();
    parent.down('hiddenfield[name=latitude]').setValue
      (position.lat());
    parent.down('hiddenfield[name=longitude]').setValue
      (position.lng());
  }
},
```

When clicked, set the map marker appropriately and attach `latitude` and `longitude` to the appropriate fields. When the form is saved, this data will be persisted.

Finally, for the `Clear` button, do the following:

```
{
  xtype: 'button',
  text: 'Clear',
```

```
hidden: true,
handler: function(button) {
  var parent = button.up('todo-map');
  parent.clearMarker(parent);
parent.down('hiddenfield[name=latitude]').setValue(null);
  parent.down('hiddenfield[name=longitude]').setValue(null);
}
}
```

When clicked, clear the map marker and nullify `latitude` and `longitude`. When the form is saved, the location will be cleared.

Wiring up the controller

This is it for `Map.js`. Now, we need to make a few changes to the `Main` controller. Our app follows the MVC pattern, which means that we need to link our views to the backing data by means of a `controller` class. Remember how the map markers are set when the map is dynamically added to its panel? Let's implement these dynamic operations now.

Edit `todo-app/app/controller/Main.js` and make the following changes to the `showView` method:

```
showView: function(view) {
  if (!this.mapResource) {
    this.mapResource = Ext.create('widget.map', {
      xtype: 'map',
      width: 'auto',
      height: 300,
      mapOptions: {
        zoom: 15
      },
      listeners: {
        maprender: function(obj, map) {
          var parent = this.up('todo-map');
          parent.onMapRender(obj, map);
        }
      }
    });
  }
  var map = this.getMainPanel().down('todo-map');
  if (map) {
    map.down('panel').remove(this.mapResource, false);
  }
  this.getMainPanel().removeAll();
```

```
    this.getMainPanel().add(view);
    map = this.getMainPanel().down('todo-map');
    if (map) {
      map.down('panel').add(this.mapResource);
    }
  },
```

Here's where the magic happens. We attach a map resource to the controller. This resource is a singleton. It is created once and added to our custom component as needed. The reason for this is that an application may only have one instance of a Google map. Add another map and the app fails to display either map correctly.

Every time the view is changed, we check for usages of our map resource. If used, we remove it from the associated panels. If the new view has a map, we add it to the panel. Due to our onMapAdd handler method, it reads the data from this component and sets the markers appropriately when the map is added to a new component:

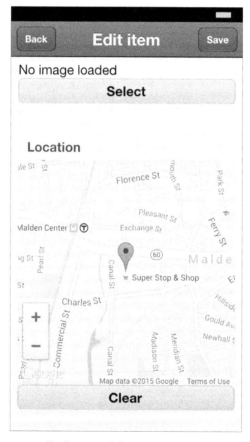

To-do app with basic map support

Refine the map.

The map works, kind of, but it's not optimal. The map is a little small; we really want it to be fullscreen. Also, scrolling feels rough as both the page and map scroll vertically at the same time. Let's fix this now by implementing a fullscreen map view, used when setting the location.

Building the map view

First, let's create this new view:

```
Ext.define('TodoApp.view.Location', {
  extend: 'Ext.Panel',
  alias: 'widget.todo-location',
  requires: [
    'Ext.Map'
  ],

  config: {
  layout: 'fit',
    items: [
      {
        docked: 'top',
        xtype: 'titlebar',
        title: 'Set location',
        items: [
          {
            align: 'left',
            text: 'Back',
            action: 'back'
          },
          {
            align: 'right',
            text: 'Set',
            action: 'set'
          }
        ]
      },
      {
        xtype: 'panel',
        layout: 'fit',
        listeners: {
```

```
            add: function(obj, item, index) {
              var parent = obj.up('todo-location');
              parent.onMapAdd(obj, item.getMap());
            }
          }
        }
      ]
    },

    onMapAdd: function(obj, map) {
      var mapPanel = obj.up('todo-location');

      // Resize the Google map
      google.maps.event.trigger(map, 'resize');

      if (map.mapRendered) {
        mapPanel.clearMarker(mapPanel);
      }
    },

    clearMarker: function(me) {
      var map = me.down('map').getMap();

      map.mapMarker.bindTo('position', map, 'center');

      me.down('map').setMapOptions({draggable: true});
    }
});
```

Next, let's add a reference to this new view in the `views` configuration of our `Main` controller:

```
views: [
  ...
  'Location',
],
```

Next, let's add a selector to the `refs` array in this controller:

```
locationPanel: {
  selector: 'todo-location',
  xtype: 'todo-location',
  autoCreate: true
}
```

As our app is now three levels deep (the list view, edit and new views, and location view), we need to rethink how we will drill down. Let's switch our main view to use a `card` layout. Each card represents another level in the hierarchy.

```
Ext.define('TodoApp.view.Main', {
  …
  config: {
    layout: 'card',
    items: { xtype: 'todo-list' }
  },
  activeIndex: 0
});
```

Adding drilldown support to the controller

Now, we need to modify the `showView` function in the `Main` controller. Instead of just swapping the views in and out, it needs to be aware of the `card` layout, preserve the cards that come before the current view, and remove the cards that fall after. To start, let's refactor some of the map logic in their own functions for simplification:

```
createMapResource: function() {
  var me = this;
  if (!this.mapResource) {
    this.mapResource = Ext.create('widget.map', {
      xtype: 'map',
      mapOptions: {
        zoom: 15,
        disableDefaultUI: true
      },
      listeners: {
      maprender: function(obj, map) {
      mapPanel = obj.up('todo-map') || obj.up('todo-location');

      map.mapMarker = new google.maps.Marker({
        map: map
      });

      // Trigger a resize when the bounds of the map change
      google.maps.event.addListener(map, 'bounds_changed',
        function() {
          var panel = obj.up('todo-map') || obj.up
            ('todo-location');
```

```
                        panel.onMapAdd(obj, map);
            });

            if (!mapPanel.mapRendered) {
              map.mapRendered = true;
              mapPanel.onMapAdd(obj, map);
            }
            }
            }
        });
    }
},
unloadMapResource: function() {
  var map = this.getMainPanel().down('map');

  if (map) {
    map.up('panel').remove(this.mapResource, false);
  }
},
loadMapResource: function() {
  var map = this.getMainPanel().getActiveItem().down('todo-map')
    || this.getMainPanel().down('todo-location');

  if (map) {
    map.down('panel').add(this.mapResource);
  }
}
```

Now, let's reference these functions in the body of the showView function and add our card management logic:

```
showView: function(view, index) {
  this.createMapResource();
  this.unloadMapResource();

  for (var i = this.getMainPanel().getItems().length - 1; i >=
    index; --i) {
    this.getMainPanel().remove(this.getMainPanel().getAt(i),
      false);
  }
  this.getMainPanel().add(view);
  this.getMainPanel().setActiveItem(index);
  this.getMainPanel().activeIndex = index;

  this.loadMapResource();
},
```

This is all well and good to show new views, but what if we just want to go back? That's a similar operation but we don't actually need to add or remove any views (besides the shared map resource). Let's add a `goBack` function now:

```
goBack: function() {
  this.unloadMapResource();

  if (this.getMainPanel().activeIndex > 0) {
    this.getMainPanel().activeIndex--;
  }
  this.getMainPanel().setActiveItem
    (this.getMainPanel().activeIndex);

  this.loadMapResource();
},
```

Wiring up the map view

With these new functions in place, let's rewire the listeners in the control object now. The back buttons should point to the `goBack` function and we have new buttons related to the `Location` view, so let's wire these up at the same time:

```
'todo-new button[action=back]': {
  tap: 'goBack'
},
'todo-edit button[action=back]': {
  tap: 'goBack'
},
'todo-map button[action=set]': {
  tap: 'showLocationView'
},
'todo-location button[action=back]': {
  tap: 'goBack'
},
'todo-location button[action=set]': {
  tap: 'setLocation'
}
```

Obviously, we haven't implemented the `showLocationView` or `setLocation` views yet, so let's do this next. The first is used to show the fullscreen map view and the second is used to take the selection and apply it to whichever view comes previously in the `card` layout (either the `todo-new` or `todo-edit` views):

```
showLocationView: function() {
  this.showView(this.getLocationPanel(), 2);
```

```
  },
  setLocation: function() {
    var panel = this.getLocationPanel(),
      position,
      map;

    position = panel.down('map').getMap().mapMarker.getPosition();
    this.goBack();

    panel = this.getMainPanel().getActiveItem().down('todo-map');
    panel.down('hiddenfield[name=latitude]').setValue
      (position.lat());
    panel.down('hiddenfield[name=longitude]').setValue
      (position.lng());
    panel.hideMap(panel, false);
    panel.setMarker(panel, position.lat(),
      position.lng());
  },
```

Finally, we need to tweak a few of the existing functions in order to adapt them to the `card` layout as well. Previously, in the process of swapping views in and out, the removed views were destroyed and then recreated on demand. This meant that we started from a clean slate each time that we opened a view. Now, the views are (mostly) persistent, which means that we need to manage certain aspects manually.

The `createTodoItem` function is now responsible for refreshing the list of to-do items when a new item is created:

```
  createTodoItem: function(button, e, eOpts) {
    var store = Ext.getStore('Item');

    store.add(this.getNewForm().getValues())
    store.sync();
    store.load();
    this.showListView();
  },
```

The `editTodoItem` function is now responsible for clearing any existing image in addition to loading them. Edit the `mediaData` block:

```
// Show the associated image
if (mediaData) {
  imagePanel.down('panel').setHtml('<img src="' +
    record.get('media') + '" alt="todo image" width="100%"/>');
  imagePanel.down('button[text=Select]').setHidden(true);
  imagePanel.down('button[text=Remove]').setHidden(false);
} else {
  imagePanel.down('panel').setHtml('No image loaded');
  imagePanel.down('button[text=Select]').setHidden(false);
  imagePanel.down('button[text=Remove]').setHidden(true);
}
```

The `showNewView` function is now responsible for resetting itself, including any displayed image whenever it is shown:

```
showNewView: function() {
  var newPanel = this.getNewPanel(),
    newForm = this.getNewForm();

  // Reset the new panel
  newForm.reset();
  newForm.down('todo-image').down('panel').setHtml('No image loaded');
  newForm.down('todo-image').down
    ('button[text=Select]').setHidden(false);
  newForm.down('todo-image').down('button[text=Remove]').
setHidden(true);

  this.showView(newPanel, 1);
},
```

Refactoring the map views

This is it for the Main controller. Now, we need to modify our Map view. As this view shares much in common with the new Location view, we've abstracted out much of the functionality and put it in the Main controller instead. The resulting view is simpler than before.

First, let's add a container right after the hidden latitude and longitude fields, which are displayed when no location has been selected (similar to the message when no image has been selected):

```
{
  xtype: 'container',
  layout: 'fit',
  html: 'No location selected'
},
```

Next, let's remove handler from the Set button and replace it with an action as this logic is now handled by the Main controller:

```
{
  xtype: 'button',
  text: 'Set',
  action: 'set'
},
```

Now, let's simplify handler for the Clear button by abstracting the logic into a separate function named hideMap. The ability to hide the map is needed at multiple points in the code, hence the reason for this:

```
{
  xtype: 'button',
  text: 'Clear',
  hidden: true,
  handler: function(button) {
    var parent = button.up('todo-map');
    parent.hideMap(parent, true);
  }
}
```

Now for the `hideMap` function as follows:

```
hideMap: function(me, hidden) {
  if (hidden) {
    me.down('hiddenfield[name=latitude]').setValue(null);
    me.down('hiddenfield[name=longitude]').setValue(null);
  }

  me.down('container').setHidden(!hidden);
  me.down('panel').setHidden(hidden);
  me.down('button[text=Set]').setHidden(!hidden);
  me.down('button[text=Clear]').setHidden(hidden);
}
```

Now you can delete the `onMapRender` and `clearMarker` functions as these functions are now handled by the `Main` controller. As we have separate views to handle the viewing and setting of the location, we don't need multiple markers attached to our Google map, just the one. This means that we can simplify the `setMarker` function as well:

```
setMarker: function(me, latitude, longitude) {
  var map = me.down('map').getMap(),
    position;

  map.mapMarker.setPosition(new google.maps.LatLng(latitude,
    longitude));

  me.down('map').setMapOptions({
    center: map.mapMarker.getPosition(),
    draggable: false
  });
},
```

Last, we need to trigger a resize of the Google map whenever its container changes its size. The `todo-map` and `todo-location` views are 300 pixels high and fullscreen, respectively, so we should trigger a resize whenever we switch between them. Add the following code to the `onMapAdd` function right after the variable initialization:

```
// Resize the Google map
google.maps.event.trigger(map, 'resize');
```

You're done! Your fullscreen map view should appear as follows:

To-do app with a fullscreen map

Comparing and contrasting with the design principles

We added two new features in this chapter: attaching an image to a to-do item and specifying a location on a map. Let's put our phone in the airplane mode and do another quick evaluation against the principles with the same pass/fail scoring as before.

Give me uninterrupted access to the content I care about.

One of the major shortcomings of our app is that the map functionality does not work offline. When you try to view or edit the map, you get a rather unfortunate error message:

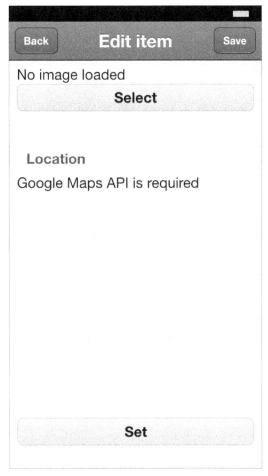

Offline map error

This error occurs because the app can't load the required Google Maps API script, which prevents us from viewing maps or changing our coordinates. The basic problem is that Google does not let you cache their maps without paying for this ability. To get around this, you can use open maps, which have no such restriction.

Unfortunately, doing this is complicated and outside the scope of this book, so I leave it as an exercise for the reader. However, to prevent this problem from affecting how we score against the design principles, I won't count it against us. Pass.

Content is mutable. Don't let my online/offline status change that.

This is a fail. Due to the error mentioned previously, it is impossible to change the map coordinates when offline because we depend on a JavaScript file that is available only online. We'll show how to cache the JavaScript (and maps) in the next chapter.

Error messages should not leave me guessing or unnecessarily worried.

This is a fail. The map error is not especially clear. It doesn't explain that the source of the problem is that I'm offline. It also doesn't tell me what to do to resolve the problem. Finally, while the message isn't particularly worrisome, it isn't very reassuring either.

Don't let me start something I can't finish.

This is a pass. You can still save changes to the to-do items while offline, even though the map doesn't work.

An app should never contradict itself. If a conflict exists, be honest about it.

This is a pass. It's bad that we don't give you any map coordinates while offline but it's good that we don't try and show you the wrong thing.

When the laws of physics prevail, choose breadth over depth when caching.

This is a fail. We don't cache the map at all so this principle is moot in a bad way. We'll introduce caching in the next chapter.

Empty states should tell me what to do next in a delightful way.

This is a fail. We still don't provide much guidance when you're filling out a to-do item for the first time.

Don't make me remember what I was doing last. Remember for me.

This is a fail. We haven't yet provided a way to keep the context sticky. We'll address this problem in the next chapter.

Degrade gracefully as the network degrades. Don't be jarring.

This is a pass. The mapping behavior is very sane as the network degrades. Once the mapping library has loaded, it displays any cached maps with no problems, regardless of the network state.

Never purge the cache unless I demand it.

This is a fail. The Maps API has a short-term cache but is cleared when the app quits. We want the maps to persist between loads of the app so that people don't have to worry about it disappearing.

Where does it need to improve?

We didn't do nearly as well this time around. Our app only meets 4/10 of the principles, down from 8/10 in the previous chapter. This is entirely due to the Google Maps API. As we rely on a library and data that is external to our app, we're technically "online" but in the most naive way possible.

In the next chapter, we'll introduce caching. This will help us improve in most of these areas. Additional improvement is needed in our error messaging, which we'll also address soon.

Summary

In this chapter, we designed the online behavior for our app, looked at state transitions and their effects on the workflow, and customized Cloudant, which we will use in the following chapters to support synchronization and list sharing. We also continued to refine our offline app by adding support for images and map locations.

In the next chapter, we will talk about the different libraries that are available for offline caching, pick one, connect our app to our web service, and improve the design and readability of the default error states. This will allow us to do another round of evaluation against the principles and set us up to continue adding functionalities in the following chapters.

4
Getting Online

We've seen that adding a little bit of online capability can really hurt our adherence to the offline design principles. By default, most libraries and frameworks don't handle being offline very gracefully, so we need to implement our own handling of these situations. In this chapter, we'll do exactly this.

Caching enables the illusion of being online when you're not. However, there are two types of caching: short-term caching, which is persistent per session, and long-term caching, which is persistent across sessions. In the context of mobile app development, restarting the app will clear the short-term cache but should leave the long-term cache intact. Most apps support short-term caching by default. To enable long-term caching, you'll need to do a bit more work.

Fortunately, solving the long-term caching problem is relatively straightforward. There are a great number of libraries and tools to make this easy. We'll discuss a few of them in this chapter, including **PouchDB**, **remoteStorage**, and **Hoodie**.

Of course, whether you have a cache or not, the data has to come from somewhere. In our to-do app, the data is persisted in the browser's local storage. This is a simple offline storage solution, adequate for most needs. However, as we need to share data with our friends and multiple devices, we need to make our data more accessible. That's why we configured a CouchDB database on Cloudant in the last chapter. We'll need to reconfigure our app to pull data from this service directly, thus making it a truly online experience.

However, sharing and syncing has its own set of problems. We'll need to support user accounts and multiple devices per account. Both of these are features that you can manage only while online, so how do we make this clear in the UI? A first attempt at getting online can create more issues than it can solve.

In this chapter, we'll investigate some options for offline database caching, connect our app to our Cloudant instance, and implement user, sharing, and notification features that only function while online. In addition, we'll shore up some of the shortcomings in the to-do app with our newfound caching knowledge, noting the error messages as we go.

By the end of this chapter, the to-do app will be fully online. The rest of the book will be spent polishing the rough edges and improving the offline experience. Let's get started.

Offline databases

There is really only one requirement of a good online database: synchronization. If a database solves the problem of making multiple databases eventually consistent, it's probably an option worth considering. There are many choices available: **derby.js**, **Lawnchair**, **Firebase**, **remotestorage.io**, **Sencha Touch**, Hoodie, PouchDB, and others.

Some of these are pure databases; others provide end-to-end offline-first development frameworks. For simplicity, we will focus on four: Sencha Touch, PouchDB, remotestorage.io, and Hoodie.

Sencha Touch

We've chosen Sencha Touch to build our offline app. As we saw in *Chapter 2, Building a To-do App*, Sencha Touch provides support for multiple types of backing stores and makes it trivial to swap between them. Some of these backing stores, such as **LocalStorage**, are offline-based. In a way, Sencha Touch is less about storing data offline and more about providing an easy interface layer between the frontend and data.

Why isn't LocalStorage sufficient? As it is restricted to a single browser, it can't be used for any kind of meaningful sharing or synchronization across devices. At least, not without help. It's too restrictive. While it makes an excellent option to build the offline nucleus of your app, it's inadequate once you need to hybridize the online and offline worlds.

Once you decide to craft a hybrid solution that functions both online and offline, what is needed? You still need a local backing store for the caching. Without LocalStorage, you've lost all of your data as soon as your app goes offline. However, you also need an authoritative remote source for the data. That's why we created a hosted CouchDB database in the last chapter. It's a place where data for multiple users and devices can be consolidated, flowing in and out as needed.

This ebb and flow implies at least two databases, a local cache and remote source of truth, which talk to one another. Synchronization between a local backing store and the remote source of truth is the critical thing. A traditional database isn't accustomed to sharing but an offline app absolutely demands sharing. It can't assume connectivity to a remote database but must have local access to the data at all times. This is why we used CloudDB because it is built from the ground up for synchronization. **MySQL**, **PostgreSQL**, and even other **NoSQL** databases such as **MongoDB** are not.

Not to say that you can't use a traditional database to build an offline-first app. It's just harder. We're going to stay with the path of least resistance in this book.

So if LocalStorage isn't adequate for a local caching database, what other local backing stores does Sencha Touch support out of the box? The only other choice is **MemoryProxy**, which isn't better as it just stores records in memory instead of in a cookie. We need to look beyond the Sencha ecosystem for other alternatives. Let's inspect our remaining options.

PouchDB

Where Sencha Touch is a frontend framework that can interface with storage, PouchDB is a backend database that can interface with a frontend framework. It's a JavaScript-based implementation of the CouchDB API with the goal of near-perfect emulation. As with CouchDB, it supports synchronization, revision history, and provides a native REST API with no additional protocols or drivers needed.

The revision control is worth an additional mention. It is similar to Git, so every update must include an attribute with the revision ID. When conflicts arise, you get to choose how to resolve them: either automatically or by letting the user decide. As revisions are never deleted, you have a complete history and can undo or view this history quite easily.

Obviously, storing history can consume a lot of space. PouchDB allows you to compact the database, which deletes the history, or delete the database altogether. However, the revision history will still exist on the remote database so you can query it, if needed.

Development tools for PouchDB are available on Chrome and Firefox. The database itself is installed and managed with npm and has only two dependencies: CouchDB and **Cross-Origin Resource Sharing** (**CORS**).

The killer feature of PouchDB is its support for live database replication. When a change is committed, all the apps with a connection to the remote database get an instant receipt of the change. This allows you to warn users when they're editing something that has changed underneath them. Alternatively, you can simply note the fact and provide an option to overwrite or undo after the change is made. When an app goes offline, database replication stops but you can configure PouchDB to resume the synchronization once the app goes online again.

This seems like exactly the kind of database that we need. It provides synchronization and will interface easily with both Sencha Touch and our **CloudDB** web service. Let's see if either of the two remaining alternatives match what PouchDB has to offer.

The remotestorage.io library

While PouchDB is clearly a straightforward option, remotestorage.io seems to cater to a different crowd. First, it bakes the concept of users into the solution. Every user has data that they own and may even decide where this data should be stored. Out of the box, remotestorage.io has support for multiple backing stores, including Dropbox, Google Drive, IndexedDB, LocalStorage, and in-memory storage. If you write multiple apps, they can share data among themselves, so you'll only need to fill out your profile once!

Unlike other NoSQL databases, it supports folders and files as first-class data types along with the more traditional **JSON** objects. It supports synchronization and sharing but not as nicely as PouchDB.

One interesting quality is that remotestorage.io has several different caching strategies: **FLUSH** where only outgoing changes are cached until they're actually submitted, **SEEN** where only the documents that have been read or written are cached, and **ALL** which is more proactive and caches anything that is not in the process of being changed.

In general, it seems that remotestorage.io is an excellent solution if you want a choice about where your data is stored and you want this data to be freely accessible from other apps. However, it does not seem like a simple solution nor quite as robust as PouchDB. We'll pass on it for now.

Hoodie

Where PouchDB and remotestorage.io are very database focused, Hoodie encompasses all aspects of an offline-first experience. `Hood.ie` is an inspirational website, so you should definitely browse it even if you decide not to use their tools. However, it's also a disorganized website, difficult to wrap your mind around or get specific usage examples. PouchDB is still better in this regard.

Hoodie has both a frontend component (**Hoodie.js**) and backend component (**hoodie-server**). There is also an interface for PouchDB, so if you don't want to use Hoodie's database, you can choose an alternative. One of the nicest perks is that Hoodie takes care of the infrastructure around user accounts, sharing, and even e-mails. This requires an account on `Hood.ie` and deploying your app to this domain.

However, as we're building our app with Sencha Touch, Hoodie is a bit more than we need. PouchDB fills the one crucial gap that we have and has excellent documentation, so it's going to be the best option for us. This said, if you're interested in an end-to-end solution to design, build, and deploy an offline app, give Hoodie a look.

Connecting to our Web API

With PouchDB as our offline database of choice, let's switch over from LocalStorage and connect it to our remote CouchDB database. We'll need to download PouchDB, add it to our Sencha Touch app, and configure it properly.

Adding PouchDB to the app

First, download PouchDB and copy it to your app folder:

1. Open `www.pouchdb.com` in your browser.
2. Click **Download** at the top of the page.
3. Copy this file (named `pouchdb-3.6.0.min.js` or similar) to the `todo-app/` folder.

Next, edit `app.json` and add this file to the `js` array:

```
{
  "path": "pouchdb-3.6.0.min.js"
}
```

Finally, add an `init` function to the `Item` store, where you'll initialize both the local and remote PouchDB databases. The local database will use LocalStorage and remote database will use the Cloudant database that we created in the previous chapter:

```
remoteDB: null,
localDB: null,
initialize: function() {
  var me = this;

  me.remoteDB = new PouchDB
    ('https://username:password@subdomain.cloudant.com/lists');
  me.localDB = new PouchDB('lists');
}
```

Note that we will not change where Sencha Touch stores its data by default. This will remain untouched. Instead, whenever records are loaded, added, removed, or updated, we will copy these changes to our local PouchDB and then sync them to the remote PouchDB. It should look as follows:

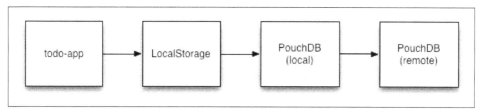

Data stores

When the app first loads, it should display whatever exists in **LocalStorage**. Next, it kicks off replication between the local and remote PouchDB instances. After this is complete, it overwrites **LocalStorage** with the contents of the **PouchDB(local)**. The advantage of this is that we always show the user something while our PouchDB instances are working in the background.

Enabling live synchronization

Normally, PouchDB will only replicate to other databases when you specifically request it. This is fine for most usages but we want to be notified when our to-do items change on another device. To support this, PouchDB provides us with a live replication mode so that changes on one device are automatically propagated to the other devices.

Add the following code to the end of the store's initialize function:

```
me.localDB.sync(me.remoteDB, {
  live: true,
  retry: true
}).on('change', function (change) {
  if (change.direction == "pull" && change.change.docs.length > 0) {
    console.log("Change occurred. Synchronizing.");
    me.load();
  }
}).on('paused', function (info) {
}).on('active', function (info) {
}).on('error', function (err) {
});
```

The live attribute enables the live replication mode while retry tells PouchDB not to error when the connection temporarily fails (such as in the offline mode). The on clauses are triggered whenever a change occurs or when synchronization pauses, resumes, or errors out. As we've enabled retry, errors should not occur but we include it for completeness.

Update PouchDB when changes occur

Now, we'll edit the configuration object in our store and add the following code to ensure that PouchDB is populated when anything changes in the underlying LocalStorage. First, let's enable autoSync so that we don't have to call sync() manually in the controllers:

```
autoSync: true,
```

Next, let's create an empty listeners block. The rest of the code in this section will live in this block:

```
listeners: {
}
```

The first event that we'll handle is load, which fires whenever the underlying store is loaded:

```
load: function(store, records, successful, operation) {
  store.localDB.get('1').then(function (doc) {
    // Add all documents to the store
    store.setData(doc.items);
  });
},
```

First, we retrieve the document that stores our list. Then, we override LocalStorage with the items array in this document.

Next, let's create an event handler. An event handler ties actions with logic. For example, `Ext.data.Store` has an action named `addrecords`, which fires whenever items are added to the store. Let's attach logic to this action now:

```
addrecords: function(store, records) {
  var toadd = [];
  // Get the document from the remote store
  store.localDB.get('1').then(function (doc) {
    for (var i = 0; i < records.length; ++i) {
      if (doc.items.every(function(e) { return e.id !=
        records[i].getData().id })) {
        toadd.push(records[i].getData());
      }
    }
    if (toadd.length > 0) {
      doc.items = doc.items.concat(toadd);
      store.localDB.put(doc);
    }
  });
},
```

As before, we start by loading our to-do list document from the local store. Next, we iterate through the items in this list. If any of the records being added don't already exist in the list, we append them to the temporary `toadd` array. If any of the records are new, we add them to the document and save it; otherwise, we do nothing.

Next, let's handle the `removerecords` event, which is fired whenever a record is deleted from the underlying store:

```
removerecords: function(store, records, indices) {
  store.localDB.get('1').then(function (doc) {
    for (var i = 0; i < records.length; ++i) {
      doc.items = doc.items.filter(function(e) { return e.id !=
        records[i].getData().id; });
    }
    store.localDB.put(doc);
  });
},
```

Similar to the previous handler, we load the document and iterate through it. However, this time, we use the `filter` function to remove all of the referenced `records` from the array and then save the document.

Finally, let's handle the `updaterecord` event, which is fired whenever an existing record is updated:

```
updaterecord: function(store, record, newIndex, oldIndex,
modifiedFieldNames, modifiedValues) {
  if (modifiedFieldNames.length == 0) {
    // No changes, don't bother updating the list
    return;
  }
  store.localDB.get('1').then(function (doc) {
    doc.items = doc.items.map(function(e) {
      if (e.id == record.getData().id) {
        return record.getData();
      } else {
        return e;
      }
    });
    store.localDB.put(doc);
  });
}
```

In this handler, we start by checking whether there's anything to change. If not, we do nothing. If there are changes to be made, we get the document as before and then update the items array, replacing any updated items with their new content.

At this point, our store fully supports our PouchDB instances. Let's make a few changes to our `Main` controller to wrap this up.

Tweaking how the store is used

We made some significant changes to how the store works. Let's adjust our controller to accommodate these changes. First, remove the calls to `sync()` and `load()` from `createTodoItem`:

```
createTodoItem: function(button, e, eOpts) {
  var store = Ext.getStore('Item');

  store.add(this.getNewForm().getValues());

  this.showListView();
},
```

As our store has enabled both `autoLoad` and `autoSync`, these calls are unnecessary. Next, replace the calls to `erase()`, `load()`, and `refresh()` with a call to `remove()` in `deleteTodoItem`:

```
deleteTodoItem: function(button, e, eOpts) {
  var dataview = this.getListDataView(),
    store = dataview.getStore(),
    record = store.findRecord('id', button.getData()).erase();

  store.remove(record);
},
```

As the data view is bound to the store, changes to the store are reflected in the data view automatically. We don't need to call `refresh()`. Additionally, to trigger our new `store` events, we need to delete `record` from the `store` directly and not via the model. This is done using the `store.remove()` function.

Sharing lists between devices

You're done! Now it's time to test your changes. First, on the command line, rebuild your changes:

```
$ sencha app build native
```

Now, open the app in your browser and on your phone. First, notice that both the instances of the app load the same list. This is because all the data is coming from a single data source: the Cloudant database that you created in the last chapter.

Second, notice that whenever you create, edit, or delete an item on one device, the other device updates the same way after a second or two. This is because of the live replication that you set up earlier. At this point, our to-do app is really online. All that's left is to start implementing the offline-only features.

Third, after you've synchronized with the online database, put your phone into airplane mode. Then, make a change in the web browser. Notice how the lists differ as the phone lacks Internet connectivity. Now, turn off the airplane mode and wait for a few seconds. The lists reconcile, which is just the behavior that we want.

Before you go any further, be sure to commit your changes.

Adding online-only features

In the last section, we connected our app to the Internet. Now, let's start to implement the online-only features, which we outlined in the previous chapter. PouchDB gives us some of this for free; namely, synchronization between devices. However, this assumes that all of the devices are linked to a single, hard-coded user account. Let's inject some flexibility into this.

First, it should be possible to manage your user account and keep your lists separate from the lists belonging to other users. Next, it should be possible to share lists between users for collaboration purposes. As with synchronization, both of these use cases are online-only so we shouldn't expose them when we're offline.

User account management

In a typical app, you need to handle user account management in the app itself. Most of these tasks—creating an account, changing passwords, deleting the account, and others—are neither relevant nor interesting in the context of offline-first app development. Consequently, we're not going to build these features. We'll use the Cloudant dashboard to manage all of this and focus on the one aspect of user account management that is very relevant to the offline user experience: signing in.

As signing in isn't something that you can do while offline, it's important that it should not prevent the user from getting value out of the app. However, once the user does sign in, we can't just discard any items that they've already created. These items must be preserved and linked to the new or existing account.

Generating credentials

Before we begin, let's create an additional user with read/write access to the lists database. You will use these credentials to sign in using the app. Open the Cloudant dashboard in your browser and do the following:

1. Choose the lists database.
2. Click **Permissions**.
3. Click **Generate API key**.
4. Save the **Password**.
5. Grant this user read and write permissions.

There are now two users with access to the lists database (in addition to the administrator). Each of these users will own a list, which they can edit independently from each other.

Expanding the stores

Once a user logs in, we should remember their credentials for the next time the app is loaded. To do this, we'll create another `model/store` pair in Ext JS, which is used to store credentials in a persistent fashion. Create `todo-app/app/model/User.js` and add the following:

```
Ext.define('TodoApp.model.User', {
  extend: 'Ext.data.Model',
  requires: [
    'Ext.data.identifier.Uuid',
    'Ext.data.proxy.LocalStorage'
  ],
  config: {
    identifier: {
      type: 'uuid'
    },
    fields: [
      'id',
      'username',
      'password'
    ],
    proxy: {
      type: 'localstorage',
      id: 'todoapp-users'
    }
  }
});
```

This model stores the current user's `username` and `password` in LocalStorage. As the proper security and authentication mechanisms are out of scope for this book, I've compromised by storing the password in clear text. For any production app, hash any passwords you store locally so that they can't be compromised.

Next, let's make the store that exposes this model to the application. Create `todo-app/app/store/User.js` and add the following:

```
Ext.define('TodoApp.store.User', {
  extend: 'Ext.data.Store',
  config: {
    model: 'TodoApp.model.User',
    autoSync: true,
    autoLoad: true
  }
});
```

Pretty simple. Note that we added the `autoSync` and `autoLoad` attributes so that we don't need to worry about calling `load()` and `sync()` manually, which will simplify our code elsewhere.

Creating the sign in view

Now, let's create a new sign in view for our app. This is where the user will enter the username and password that we've generated for them in the Cloudant dashboard. Create `todo-app/app/view/SignIn.js` and add the following:

```
Ext.define('TodoApp.view.SignIn', {
  extend: 'Ext.Panel',
  alias: 'widget.todo-sign-in',
  requires: [
    'Ext.TitleBar',
    'Ext.field.Email',
    'Ext.field.Password'
  ],

  config: {
    items: [
      {
        docked: 'top',
        xtype: 'titlebar',
        title: 'Sign in',
        items: {
          align: 'left',
          text: 'Back',
          action: 'back'
        },
      },
      {
        xtype: 'formpanel',
        scrollable: false,
        height: '100%',
        items: [
          {
            xtype: 'emailfield',
            name: 'username',
            label: 'Username'
          },
          {
            xtype: 'passwordfield',
            name: 'password',
```

```
                label: 'Password'
            },
            {
                xtype: 'button',
                text: 'Sign in',
                action: 'submit'
            }
        ]
    }
    ]
    }
});
```

You should recognize the view as being similar to the others in our app. It has a toolbar at the top with a `back` button and a form with an `emailfield`, `passwordfield`, and `submit` button. This view is accessed from the list view, so let's edit `todo-app/app/view/List.js` now. Change the `titlebar items` attribute to an array and add the following to the beginning of this array:

```
{
    align: 'left',
    text: 'Sign in',
    action: 'signin'
},
{
    align: 'left',
    text: 'Sign out',
    action: 'signout',
    hidden: true
},
```

By default, the `Sign in` button is shown while the `Sign out` button is hidden. Once a connection is made, this is swapped. Only one button is visible at a time.

Updating the controller references

Now, let's make a variety of changes to our `Main` controller. Edit `todo-app/app/controller/Main.js` and reference the new `SignIn` view from the views configuration:

```
'SignIn',
```

Now, add the User model and User store to the models and stores configurations, respectively:

```
models: [
    'User',
    'Item'
],
stores: [
    'User',
    'Item'
],
```

Next, create new selectors in the refs configuration:

```
signInPanel: {
    selector: 'todo-sign-in',
    xtype: 'todo-sign-in',
    autoCreate: true
},
signInForm: 'todo-sign-in formpanel'
```

The last configuration item that we need to change is the control configuration. Add the following listeners:

```
'todo-list button[action=signin]': {
    tap: 'showSignInView'
},
'todo-list button[action=signout]': {
    tap: 'signOut'
},
'todo-sign-in button[action=back]': {
    tap: 'goBack'
},
'todo-sign-in button[action=submit]': {
    tap: 'signIn'
}
```

All that's left to do is add controller functions to implement the functions that we've referenced here. First, let's add a single instance to this controller right after the config hash:

```
syncHandler: null,
```

This variable is very important. It contains a reference to any sync operation in progress. When we sign out, we'll call cancel() on this variable to end the sync and reset our databases in a graceful fashion.

Adding the handler functions

Immediately after this variable, define an `init` function for the controller:

```
init: function() {
  var me = this,
    store = Ext.getStore('Item'),
    record = Ext.getStore('User').first(),
    data;

  store.localDB = new PouchDB('lists');

  if (record) {
    data = record.getData();
    store.username = data.username;
    me.connect(data.username, data.password);
  }
},
```

When your app starts, this function initializes our local `PouchDB` database and checks to see if the user credentials are already present. If they are, it connects to the remote database.

Next, add the `showSignInView` function:

```
showSignInView: function() {
  this.showView(this.getSignInPanel(), 1);
},
```

When the **Sign in** button is clicked, this button shows the appropriate view. Now, let's wire up the `signIn` button in this view. Add the following function:

```
signIn: function() {
  var values = this.getSignInForm().getValues();
  this.getSignInForm().down('passwordfield').reset();
  this.connect(values.username, values.password);
  this.showListView();
},
```

This method retrieves the credentials from the form, resets the `passwordfield`, and attempts to connect to the remote database. It then returns to the `list` view. We'll implement the `connect` function in a bit but first, let's wire up the `signOut` button:

```
signOut: function() {
  this.disconnect();
},
```

This method is very simple. It merely undoes what `connect()` sets up. Now, let's implement the final methods in the controller: `connect()`, `disconnect()`, and `startSyncing()`. First, connect the following:

```
connect: function(username, password) {
  var itemStore = Ext.getStore('Item'),
    userStore = Ext.getStore('User');

  userStore.removeAll();
  userStore.add({
    username: username,
    password: password
  });

  itemStore.username = username;
  itemStore.password = password;

  if (this.syncHandler) {
    this.syncHandler.cancel();
  } else {
    this.startSyncing(this);
  }
},
```

This method takes `username` and `password` and adds these credentials to `userStore` (removing any credentials already in the store). It then adds the credentials as instance variables to `itemStore` for reference by the methods in that class. Finally, it starts the sync operation, first calling `cancel` on an existing sync operation, if necessary.

Now, let's implement `disconnect()`:

```
disconnect: function() {
  var itemStore = Ext.getStore('Item'),
    userStore = Ext.getStore('User');

  if (this.syncHandler) {
    userStore.removeAll();
    itemStore.removeAll();
    itemStore.username = 'nobody';
    itemStore.password = null;
    this.syncHandler.cancel();
  }
},
```

If a sync operation is in progress, it removes all the items from both `userStore` and `itemStores`. This accomplishes two things. First, it prevents the user from being automatically signed in when the app is refreshed. Second, it ensures that the next person to use the app will not have access to items belonging to someone else.

Finally, the `disconnect()` method resets the temporary credentials in `itemStore` and cancels the sync operation. Now, let's create the `startSyncing()` method:

```
startSyncing: function(me) {
  var me = this,
  store = Ext.getStore('Item');

  store.remoteDB = new PouchDB('https://' + store.username + ':' +
    store.password + '@djsauble.cloudant.com/lists');
  me.syncHandler = store.localDB.sync(store.remoteDB, {
    live: true,
    retry: true
  }).on('change', function (change) {
    if (change.direction == "pull" && change.change.docs.length >
      0) {
      console.log("Change occurred. Synchronizing.");
      store.load();
    }
  }).on('paused', function (info) {
  }).on('active', function (info) {
  }).on('error', function (err) {
  });
  me.syncHandler.on('complete', function (info) {
    store.localDB.destroy().then(function() {
      store.localDB = new PouchDB('lists');
      me.syncHandler = null;
      me.getListPanel().down('button[action=signin]').show();
      me.getListPanel().down('button[action=signout]').hide();
      if (store.username && store.password) {
        me.startSyncing(me);
      }
    }
  });
..});
  setTimeout(function() {
    me.getListPanel().down('button[action=signin]').hide();
    me.getListPanel().down('button[action=signout]').show();
  }, 50);
}
```

The first thing that we do is connect to the remote database using the credentials provided. Next, we start live syncing while storing a handle to the operation in `syncHandler`. Next, we handle the `'complete'` event of `syncHandler`, which allows us to do a cleanup once the operation is canceled. Finally, we show the `signout` button (and hide the `signin` button) after a brief delay, which gives the page enough time to render.

Tweaking the item store

We're almost done. The last thing that we need to do is make a few changes to our `Item` store. As we've moved the store initialization logic to the controller, it's no longer needed here. Edit `todo-app/app/store/Item.js` and delete the initialize function.

Finally, we need to ensure that the list document has been created for each user when they need it. Previously, we assumed its existence but this is no longer something that we can take for granted. Add the following function to the `Item` store:

```
doWithDoc: function(func) {
  var me = this;

  me.localDB.get(me.username + '_1', function (error, doc) {
    if (error) {
      // Document doesn't exist yet. Create it.
      me.localDB.put({
        '_id': me.username + '_1',
        'name': 'list',
        'owner': me.username,
        'items': []
      }).then(function() {
        me.localDB.get(me.username + '_1', function (error, doc) {
          console.log(doc);
          func(doc);
        });
      });
    } else {
      func(doc);
    }
  });
}
```

This method first attempts to retrieve the document. If this fails, we create a new, empty document, get it, and then pass it to the provided function as an argument. If the document already exists, we also pass it as an argument. Now, we can replace the existing `get` logic in each of our event handlers with a reference to this function. In `load`, `addrecords`, and `removerecords`, we have the following line:

```
store.localDB.get('1').then(function (doc) {
```

Replace this with the following line:

```
this.doWithDoc(function(doc) {
```

Then, to ensure that the `username` variable is always set, add an additional instance variable to the store immediately after the `remoteDB` and `localDB` variables:

```
username: 'nobody',
```

You're done! Using the credentials that you created earlier, you should be able to log in to each account, add items, and observe that the items are restricted to that user's list alone. Additionally, if you add items to the list before signing in, these items are appended to any existing list once you sign in.

Commit your changes.

Multiple lists

Now that we have support for multiple user accounts, we can expand the app to support multiple lists per user account. This will become important in the next section when we implement sharing. People want to keep their private and public lives separate.

Previously, we dealt with a single list; now we're dealing with multiple lists per account. At a minimum, we'll need a way to see all of the lists associated with our account and a way to create new lists. Also, we'll need to tweak the existing views slightly in order to accommodate the expanded workflow.

Unlike the other additions in this chapter, the ability to manage multiple lists should be available while offline. Thus, we'll need to consider the offline/online transition as we did in the last section.

Refactoring our views

In an ideal world, we'd write code perfectly to start with and never have to change it. Unfortunately, this is not the case. Code is imperfect and changes constantly. To accommodate this reality, we engage in periodic refactoring. Refactoring is a practice where you move code around or restructure it to make it easier to modify and extend in the future.

In our app, we have a lot of views under `todo-app/app/view/`. Let's break these into a couple of categories to make them easier to manage. First, switch to the `view` directory and create two subdirectories:

```
$ cd todo-app/app/view/
$ mkdir item
$ mkdir list
```

Now, move the views related to items and views related to lists into `item/` and `list/`, respectively:

```
$ git mv Edit.js Image.js Location.js Map.js New.js DataItem.js item/
$ git mv List.js list/
```

Next, edit the `Main` controller and change the class references under the views attribute from relative to absolute paths:

```
views: [
  'SignIn',
  'TodoApp.view.list.List',
  'TodoApp.view.item.New',
  'TodoApp.view.item.Edit',
  'TodoApp.view.item.Location',
  'TodoApp.view.item.DataItem'
],
```

Then, open the file for each view and modify the `Ext.define()` statement to match these absolute paths.

Implementing new views

Now, you're ready to create the new list views with their supporting models and stores. Start by creating `todo-app/app/view/list/Lists.js`, the view that displays all of the lists owned by the current user:

```
Ext.define('TodoApp.view.list.Lists', {
  extend: 'Ext.Panel',
  alias: 'widget.todo-lists',
```

```
requires: [
  'Ext.TitleBar',
  'Ext.dataview.DataView'
],

config: {
  items: [
    {
      docked: 'top',
      xtype: 'titlebar',
      title: 'My Lists',
      items: [
        {
          align: 'left',
          text: 'Sign in',
          action: 'signin'
        },
        {
          align: 'left',
          text: 'Sign out',
          action: 'signout',
          hidden: true
        },
        {
          align: 'right',
          text: 'Add',
          action: 'new'
        }
      ]
    },
    {
      xtype: 'dataview',
      height: '100%',
      useComponents: true,
      defaultType: 'todo-list-dataitem',
      store: 'List'
    }
  ]
},

initialize: function() {
  // Autoload appears to be broken for dataviews
```

```
    Ext.getStore('List').load();

    this.callParent();
  }
});
```

This view is very similar to the `List` view except that it uses a different store and data item view. Note that we moved the `signin`/`signout` buttons here as this will become the view that people see by default when they open the app.

Let's make these changes now. Edit `todo-app/app/view/Main.js` and change the item in the card layout from `todo-list` to `todo-lists`. Next, edit `todo-app/app/view/List.js` and replace the `signin`/`signout` buttons with a `back` button:

```
{
  align: 'left',
  text: 'Back',
  action: 'back'
},
```

Finally, create the new data item for the Lists view. You'll notice that this is very similar to the data item for the `Item` view. Create a copy of `todo-app/app/view/item/DataItem.js` and put it in the `list/` folder. Now, make the following changes:

- Change the `Ext.define` name to `TodoApp.view.list.DataItem`
- Change `alias` to `widget.todo-list-dataitem`
- In `config`, replace the description, `edit`, and destroy items with the following code:

```
name: {
  flex: 1
},
edit: {
  text: 'Edit',
  action: 'edit',
  margin: '0 7px 0 0'
},
destroy: {
  text: 'Delete',
  action: 'delete'
},
```

- Now, replace the contents of the `dataMap` object with the following code:

```
getName: {
  setHtml: 'name'
},
getEdit: {
  setData: '_id'
},
getDestroy: {
  setData: '_id'
}
```

- Delete all of the other functions in the file and replace them with the following:

```
applyName: function(config) {
  return Ext.factory(config, 'Ext.Label', this.getName());
},
updateName: function(newName, oldName) {
  if (newName) {
    this.add(newName);
  }
  if (oldName) {
    this.remove(oldName);
  }
},
applyEdit: function(config) {
  return Ext.factory(config, 'Ext.Button', this.getEdit());
},
updateEdit: function(newButton, oldButton) {
  if (newButton) {
    this.add(newButton);
  }
  if (oldButton) {
    this.remove(oldButton);
  }
},
applyDestroy: function(config) {
  return Ext.factory(config, 'Ext.Button', this.getDestroy());
},
updateDestroy: function(newButton, oldButton) {
  if (newButton) {
    this.add(newButton);
  }
  if (oldButton) {
    this.remove(oldButton);
  }
}
```

The last new view that we need is when the user creates a new list. As with the others, this view is very similar to the existing New view. Duplicate todo-app/app/view/item/New.js and copy it to list/, then make the following changes:

- Change the Ext.define name to TodoApp.view.list.New
- Change alias to widget.todo-list-new
- Change title of the title bar to Add list
- Change title of the form panel to Name
- Replace the value of the items array with the following code:

```
{
  xtype: 'textfield',
  name: 'name'
}
```

Creating the backing store

With the bones of the new views in place, let's create the new List store, used by the Lists view that we just created. Open todo-app/app/store/List.js and replace the contents with the following:

```
Ext.define('TodoApp.store.List', {
  extend: 'Ext.data.Store',
  config: {
    model: 'TodoApp.model.List',
    autoSync: true,
    listeners: {
      load: 'onLoad',
      addrecords: 'onAddRecords',
      removerecords: 'onRemoveRecords',
      updaterecord: 'onUpdateRecord'
    }
  },
}
```

This should look very similar to the Item store. The main difference is that we've added a load event handler in order to populate the store. Next, let's add a few instance variables after the configuration block:

```
remoteDB: null,
localDB: null,
username: 'nobody',
password: null,
currentListId: null,
```

These variables point to our PouchDB instances, the current user's credentials, and current list's ID. Next, let's define a few instance methods, starting with `doWithDoc`:

```
doWithDoc: function(func) {
  var me = this;

  me.localDB.get(me.currentListId, function (error, doc) {
    if (error) {
      // Document doesn't exist yet. Create it.
      me.localDB.put({
        '_id': me.currentListId,
        'name': 'list',
        'owner': me.username,
        'items': []
      }).then(function() {
        me.localDB.get(me.currentListId, function (error, doc) {
          func(doc);
        });
      });
    } else {
      func(doc);
    }
  });
},
```

This method takes the current list ID, creates the list if it doesn't already exist, and then passes the contents of the list to the specified callback. Next, let's define the `doWithDocs` method:

```
doWithDocs: function(func) {
  var me = this;

  me.localDB.allDocs({
    include_docs: true,
    attachments: true,
    startkey: '_design/\uffff',
  }, function (error, result) {
    func(result.rows.filter(
      function(e) {
        if (e.doc.owner == me.username) {

          return true;
        }
        if (!e.doc.collaborators) {
          return false;
```

```
        }
        return e.doc.collaborators.some(
          function(c) {
            return c == me.username;
          }
        );
      }
    ).map(function(e) {
      return e.doc;
    }));
  });
},
```

This method is similar but instead takes the contents of all the lists that the current user owns or collaborates with and passes them to the specified callback. With these helper methods out of the way, let's implement the event handlers.

First, create the onLoad method:

```
onLoad: function(store, records, successful, operation) {
  this.doWithDocs(function(lists) {
    var toadd = [];
    for (var i = 0; i < records.length; ++i) {
      var data = records[i].getData();
      if (lists.every(function(l) { return l._id != data._id })) {
        var model = new TodoApp.model.List({
          _id: data._id.replace(/.*_/, store.username + "_"),
          name: data.name,
          owner: store.username,
          items: data.items
        });
        toadd.push(model);
      }
    }
    if (toadd.length > 0) {
      lists = lists.concat(toadd);
    }
    store.setData(lists);
  });
},
```

Whenever the list store is loaded, this method is called and it refreshes the store based on the contents of our PouchDB database. It is not a destructive method. If any lists already exist in the store, they are preserved and added to PouchDB.

Next, create the `onAddRecords` method:

```
onAddRecords: function(store, records) {
  this.doWithDocs(function(lists) {
    var toadd = [];
    for (var i = 0; i < records.length; ++i) {
      if (lists.every(function(l) { return l._id !=
        records[i].getData()._id })) {
        toadd.push(records[i].getData());
      }
    }
    if (toadd.length > 0) {
      lists = lists.concat(toadd);
      store.localDB.bulkDocs(toadd);
    }
  });
},
```

Whenever a list is added to the store, this method is responsible for adding it to PouchDB. The sister method is `onRemoveRecords`:

```
onRemoveRecords: function(store, records, indices) {
  this.doWithDocs(function(lists) {
    for (var i = 0; i < records.length; ++i) {
      lists = lists.filter(function(e) { return e._id ==
        records[i].getData()._id; });
    }
    for (var i = 0; i < lists.length; ++i) {
      lists[i]._deleted = true;
    }
    if (lists.length > 0) {
      store.localDB.bulkDocs(lists);
    }
  });
},
```

This method does the opposite. Whenever the lists are removed, this method is responsible for clearing them from PouchDB.

Finally, let's define the `onUpdateRecord` method:

```
onUpdateRecord: function(store, record, newIndex, oldIndex,
modifiedFieldNames, modifiedValues) {
  if (modifiedFieldNames.length == 0) {
    // No changes, don't bother updating the list
    return;
```

```
    }
    this.doWithDoc(function(doc) {
      var newDoc = record.getData();
      newDoc._rev = doc._rev;
      store.localDB.put(newDoc);
    });
  }
```

This method handles the case when an existing list is edited. It retrieves the old list from PouchDB, updates the fields, and saves the changes.

You should notice a lot of similarities with the `Item` store. The primary difference is that this store maps documents directly to our remote Cloudant database. As a result, the `Item` store no longer needs a load handler and its other handlers are merely proxy requests to the `List` store.

Removing sync logic from the item store

Edit `todo-app/app/store/Item.js` and replace its content with the following:

```
Ext.define('TodoApp.store.Item', {
  extend: 'Ext.data.Store',
  config: {
    model: 'TodoApp.model.Item',
    autoSync: true,
    listeners: {
      addrecords: 'onAddRecords',
      removerecords: 'onRemoveRecords',
      updaterecord: 'onUpdateRecord'
    }
  },
  currentListStore: null,
  currentListRecord: null,
  onAddRecords: function(store, records) {
    if (store.currentListStore && store.currentListRecord) {
      var data = store.currentListRecord.getData(),
      toadd = [];
      if (!data.items) {
        data.items = [];
      }
      for (var i = 0; i < records.length; ++i) {
        if (data.items.every(function(e) { return e.id !=
          records[i].getData().id })) {
          toadd.push(records[i].getData());
        }
```

```
          if (toadd.length > 0) {
            data.items = data.items.concat(toadd);
            store.currentListRecord.setData(data);
            store.currentListRecord.setDirty();
            store.currentListStore.sync();
          }
        }
      }
    },
    onRemoveRecords: function(store, records, indices) {
      if (store.currentListStore && store.currentListRecord) {
        var data = store.currentListRecord.getData();
        for (var i = 0; i < records.length; ++i) {
          data.items = data.items.filter(function(e)
            { return e.id != records[i].getData().id; });
        }
        store.currentListRecord.setData(data);
        store.currentListRecord.setDirty();
        store.currentListStore.sync();
      }
    },
    onUpdateRecord: function(store, record, newIndex, oldIndex,
      modifiedFieldNames, modifiedValues) {
      if (modifiedFieldNames.length == 0) {
        // No changes, don't bother updating the list
        return;
      }
      if (store.currentListStore && store.currentListRecord) {
        var data = store.currentListRecord.getData();
        data.items = data.items.map(function(e) {
          if (e.id == record.getData().id) {
            return record.getData();
          } else {
            return e;
          }
        });
        store.currentListRecord.setData(data);
        store.currentListRecord.setDirty();
        store.currentListStore.sync();
      }
    }
  }
});
```

This store contains all of the items for the currently selected `List` model. When an item is added, removed, or updated, we retrieve the parent model and update the items array.

Giving the list store a model

The `Item` model itself doesn't change at all but we now need a model for the list store. Create `todo-app/app/model/List.js` and add the following:

```
Ext.define('TodoApp.model.List', {
  extend: 'Ext.data.Model',
  requires: [
    'Ext.data.identifier.Uuid',
    'Ext.data.proxy.LocalStorage'
  ],
  config: {
    identifier: {
      type: 'uuid'
    },
    idProperty: '_id',
    fields: [
      '_id',
      '_rev',
      'name',
      'owner',
      'collaborators',
      'items'
    ],
    proxy: {
      type: 'localstorage',
      id: 'todoapp-lists'
    }
  }
});
```

Adding logic to the controller

The last file that we need to change is the `Main` controller. Open it now and add the new `List` model and `List` store to the relevant arrays. Next, add selectors to the `refs` array:

```
listsPanel: {
  selector: 'todo-lists',
  xtype: 'todo-lists',
```

```
    autoCreate: true
  },
  listsDataView: 'todo-lists dataview',
  newListPanel: {
    selector: 'todo-list-new',
    xtype: 'todo-list-new',
    autoCreate: true
  },
  newListForm: 'todo-list-new formpanel',
```

Next, edit the control configuration and change the `signin`/`signout` handlers to reflect their move from `todo-list` to `todo-lists`:

```
'todo-lists button[action=signin]': {
  tap: 'showSignInView'
},
'todo-lists button[action=signout]': {
  tap: 'signOut'
},
```

Add additional handlers for the `todo-lists` and `todo-list-new` views:

```
'todo-lists button[action=new]': {
  tap: 'showNewListView'
},
'todo-lists button[action=edit]': {
  tap: 'editList'
},
'todo-lists button[action=delete]': {
  tap: 'deleteList'
},
'todo-list-new button[action=back]': {
  tap: 'showListsView'
},
'todo-list-new button[action=create]': {
  tap: 'createList',
},
```

Due to the shuffling that we've done, many of the methods that point to the `Item` store should now point to the `List` store. Change this for the `init`, `connect`, `disconnect`, and `startSyncing` methods.

Additionally, as the `todo-lists` view is now the first view that people see, all of the other views are offset by one. Add one to `showView` in the following methods: `showNewView()`, `showEditView()`, `showListView()`, and `showLocationView()`.

Now, let's implement these new handler functions that we defined in the preceding control configuration. First, let's write the method responsible for creating a new list:

```
createList: function(button, e, eOpts) {
  var store = Ext.getStore('List');

  var model = new TodoApp.model.List(this.getNewListForm().
getValues());
  model.setId(store.username + '_' + model.getId());
  model.set('owner', store.username);
  store.add(model);

  this.showListsView();
},
```

Next, the method responsible for setting a particular list as the current list (and updating the Item store appropriately) is as follows:

```
editList: function(button, e, eOpts) {
  var listStore = Ext.getStore('List'),
  record = listStore.findRecord('_id', button.getData()),
  items = record.getData().items,
  listPanel = this.getListPanel(),
  listDataView = this.getListDataView(),
  itemStore = Ext.getStore('Item');

  listStore.currentListId = button.getData();
  itemStore.currentListStore = listStore;
  itemStore.currentListRecord = record;
  itemStore.removeAll();
  if (items) {
    itemStore.add(items);
  }

  listPanel.down('titlebar').setTitle(record.get('name'));
  this.showListView();
},
```

Next, the method that deletes an existing list is as follows:

```
deleteTodoItem: function(button, e, eOpts) {
  var dataview = this.getListDataView(),
  store = dataview.getStore(),
  record = store.findRecord('id', button.getData()).erase();

  store.remove(record);
},
```

Next, here is the method that shows the `Lists` view:

```
showListsView: function() {
  this.showView(this.getListsPanel(), 0);
},
```

Finally, the method that shows the `New` view to create a new list is as follows:

```
showNewListView: function() {
  var newListPanel = this.getNewListPanel(),
  newListForm = this.getNewListForm();

  // Reset the new panel
  newListForm.reset();

  this.showView(newListPanel, 1);
},
```

We're almost done. As we moved sign in to the Lists view, we should return to this view after signing in. Edit the `signIn()` method and change `this.showListView()` to `this.showListsView()`.

You're done! Commit your changes and then start the server and try signing in to the accounts that you created earlier. It should just work. Note that if you create lists without being signed in, these lists will automatically migrate to the first account that you sign in to. Pretty neat, huh?

Sharing lists

At this point, we have a bunch of different user accounts floating around, each with a number of lists tied to the account. However, they're all isolated from one another. We should fix this in such a way that people who want to share lists can do so. This implies that we need a way for people to choose a list and decide who to share it with.

Creating share views

Let's start by creating a new folder under `todo-app/app/view/` named `collaborator/`. We'll put the views associated with sharing in this folder. As these views are very similar to the existing views, we'll just duplicate the existing views and make some minor modifications.

First, create a copy of `todo-app/app/view/item/DataItem.js` and put it in the `collaborator/` folder. Now, make the following changes:

- Change the `Ext.define` name to `TodoApp.view.collaborator.DataItem`
- Change `alias` to `widget.todo-collaborator-dataitem`
- In `config`, replace the `description`, `edit`, and `destroy` items with the following code:

```
name: {
  flex: 1
},
unshare: {
  text: 'Unshare',
  action: 'delete'
},
```

- Now, replace the contents of the `dataMap` object with the following code:

```
getName: {
  setHtml: 'id'
},
getUnshare: {
  setData: 'id'
}
```

- Delete all of the other functions in the file and replace them with the following:

```
applyName: function(config) {
  return Ext.factory(config, 'Ext.Label', this.getName());
},
updateName: function(newName, oldName) {
  if (newName) {
    this.add(newName);
  }
  if (oldName) {
    this.remove(oldName);
  }
},
applyUnshare: function(config) {
  return Ext.factory(config, 'Ext.Button',
    this.getUnshare());
},
updateUnshare: function(newButton, oldButton) {
  if (newButton) {
    this.add(newButton);
  }
```

```
        if (oldButton) {
          this.remove(oldButton);
        }
    }
}
```

Now, make a copy of `todo-app/app/view/list/List.js`, move it to the
`collaborator/` folder, and make the following changes:

- Change the `Ext.define` name to `TodoApp.view.collaborator.List`
- Change `alias` to `widget.todo-collaborator-list`
- Change `title` of the title bar to `Users`
- Change `action` of the Add button to `add`
- Change `defaultType` of the data view to `todo-collaborator-dataitem`
- Change `store` of the data view to `Collaborator`

Finally, duplicate `todo-app/app/view/list/New.js`, copy it to `collaborator/`,
and make the following changes:

- Change the `Ext.define` name to `TodoApp.view.collaborator.New`
- Change `alias` to `widget.todo-collaborator-new`
- Change `name` of the create button to `Share`
- Change `action` of the create button to `share`
- Change `title` of the title bar to `Add user`
- Change `title` of the form panel to `Name`
- Change `name` of the text field to `name`

Modifying the existing views

With these new views in place, let's turn our attention to a few of the existing views.
They need some modifications in order to work with these new views. First, edit
`todo-app/app/view/list/DataItem.js` and make the following changes:

- Add the following code under the `name` object in `config`:

```
share: {
  text: 'Share',
  action: 'share',
  margin: '0 7px 0 0'
},
```

- Add the following code under `getShare` in the `dataMap` object:

```
getShare: {
  setData: '_id'
},
```

- Finally, add the following two functions to the component:

```
applyShare: function(config) {
  return Ext.factory(config, 'Ext.Button', this.getShare());
},
updateShare: function(newButton, oldButton) {
  if (newButton) {
    this.add(newButton);
  }
  if (oldButton) {
    this.remove(oldButton);
  }
},
```

Adding a model

Now, we need a new model and store to handle local persistence of the data backing these new views. A model is the M part of **Model View Controller** (**MVC**). It defines the data attributes and describes where this data is stored. First, create `todo-app/app/model/Collaborator.js` and add the following:

```
Ext.define('TodoApp.model.Collaborator', {
  extend: 'Ext.data.Model',
  requires: [
    'Ext.data.identifier.Uuid',
    'Ext.data.proxy.LocalStorage'
  ],
  config: {
    identifier: {
      type: 'uuid'
    },
    fields: [
      'id'
    ],
    proxy: {
      type: 'localstorage',
      id: 'todoapp-shares'
    }
  }
});
```

Adding a store

Let's make the store for this model. Create `todo-app/app/store/Collaborator.js` and add the following:

```javascript
Ext.define('TodoApp.store.Collaborator', {
  extend: 'Ext.data.Store',
  config: {
    model: 'TodoApp.model.Collaborator',
    autoSync: true,
    listeners: {
      addrecords: 'onAddRecords',
      removerecords: 'onRemoveRecords',
      updaterecord: 'onUpdateRecord'
    }
  },
  currentListStore: null,
  currentListRecord: null,
  onAddRecords: function(store, records) {
    if (store.currentListStore && store.currentListRecord) {
      var data = store.currentListRecord.getData(),
      toadd = [];
      if (!data.collaborators) {
        data.collaborators = [];
      }
      for (var i = 0; i < records.length; ++i) {
        if (data.collaborators.every(function(id) { return id !=
          records[i].getData().id })) {
          toadd.push(records[i].getData().id);
        }
        if (toadd.length > 0) {
          data.collaborators =
            data.collaborators.concat(toadd);
          store.currentListRecord.setData(data);
          store.currentListRecord.setDirty();
          store.currentListStore.sync();
        }
      }
    }
  },
  onRemoveRecords: function(store, records, indices) {
    if (store.currentListStore && store.currentListRecord) {
      var data = store.currentListRecord.getData();
      for (var i = 0; i < records.length; ++i) {
```

```
                data.collaborators = data.collaborators.filter
                    (function(id)
                        { return id != records[i].getData().id; });
            }
            store.currentListRecord.setData(data);
            store.currentListRecord.setDirty();
            store.currentListStore.sync();
        }
    },
    onUpdateRecord: function(store, record, newIndex, oldIndex,
        modifiedFieldNames, modifiedValues) {
        if (modifiedFieldNames.length == 0) {
            // No changes, don't bother updating the list
            return;
        }
        if (store.currentListStore && store.currentListRecord) {
            var data = store.currentListRecord.getData();
            data.collaborators = data.items.map(function(id) {
                if (id == record.getData().id) {
                    return record.getData().id;
                } else {
                    return id;
                }
            });
            store.currentListRecord.setData(data);
            store.currentListRecord.setDirty();
            store.currentListStore.sync();
        }
    }
});
```

Modifying the list store

Finally, we need to make a few changes to our list store. Previously, we were able to load just the documents associated with the current user. Now, we need to load all the lists and filter out all the lists for which the current user is neither the owner or a collaborator.

Edit todo-app/app/store/List.js and replace the doWithDocs function with the following:

```
doWithDocs: function(func) {
    var me = this;

    me.localDB.allDocs({
```

```
    include_docs: true,
    attachments: true,
    startkey: '_design/\uffff',
}, function (error, result) {
  func(result.rows.filter(
    function(e) {
      if (e.doc.owner == me.username) {
        return true;
      }
      if (!e.doc.collaborators) {
        return false;
      }
      return e.doc.collaborators.some(
        function(c) {
          return c == me.username;
        }
      );
    }
  ).map(function(e) {
    return e.doc;
  }));
});
},
```

Adding logic to the controller

We're almost done. The last thing to do is wiring up the behavior in our `Main` controller. Edit `todo-app/app/controller/Main.js` and make the following changes:

- Add references for our new views to the views configuration:

  `'TodoApp.view.collaborator.List',`

  `'TodoApp.view.collaborator.New',`

  `'TodoApp.view.collaborator.DataItem'`

- Add a reference to the `collaborator` model and store to the models and stores configurations, respectively.

- Add references for the panels and forms to the `refs` configuration:

  ```
  collaboratorsPanel: {
    selector: 'todo-collaborator-list',
    xtype: 'todo-collaborator-list',
    autoCreate: true
  },
  collaboratorsDataView: 'todo-collaborator-list dataview',
  ```

```
newCollaboratorPanel: {
  selector: 'todo-collaborator-new',
  xtype: 'todo-collaborator-new',
  autoCreate: true
},
newCollaboratorForm: 'todo-collaborator-new formpanel',
```

- Add handlers to the `control` configuration:

```
'todo-lists button[action=share]': {
  tap: 'shareList',
},
'todo-collaborator-list button[action=back]': {
  tap: 'goBack'
},
'todo-collaborator-list button[action=add]': {
  tap: 'showNewCollaboratorView'
},
'todo-collaborator-list button[action=delete]': {
  tap: 'deleteCollaborator'
},
'todo-collaborator-new button[action=back]': {
  tap: 'goback'
},
'todo-collaborator-new button[action=share]': {
  tap: 'createCollaborator'
},
```

- Add a new function to initialize the collaborator list. To do this, copy the `editList` function and make the following changes:

    1. Change the name of the function to `editList`.

    2. Set the `items` variable to `record.getData().collaborators`.

    3. Set the `listPanel` variable to `this.getCollaboratorsPanel()`.

    4. Set the `listDataView` variable to `this.getCollaboratorsDataView()`.

    5. Set the `itemStore` variable to `Ext.getStore('Collaborator')`.

6. Finally, because we need to transform the list of collaborators to a list of objects, set the argument of `itemStore.add` to the following code:

```
collaborators.map(
  function(c) {
    return {id: c}
  }
)
```

7. Add another function that creates a new collaborator based on the values in the related form:

```
createCollaborator: function(button, e, eOpts) {
  var store = Ext.getStore('Collaborator');

  store.add(this.getNewCollaboratorForm().getValues());

  this.showCollaboratorsView();
},
```

8. Now, create a function to delete the collaborators, as needed:

```
deleteCollaborator: function(button, e, eOpts) {
  var dataview = this.getCollaboratorsDataView(),
    store = dataview.getStore(),
    record = store.findRecord('id',
      button.getData()).erase();

  store.remove(record);
},
```

9. Finally, we need two trivial functions that show the collaborator list and new collaborator views:

```
showCollaboratorsView: function() {
  this.showView(this.getCollaboratorsPanel(), 1);
},
showNewCollaboratorView: function() {
  this.showView(this.getNewCollaboratorPanel(), 2);
},
```

At this point, start the application, choose a list, and share it with a collaborator. Use one of the other credentials that you created as the username. When you log in as this user, you should be able to view and edit the list that was shared with you as well as add additional collaborators:

Sharing with collaborators

There are a lot of rough edges here but we've taken our previous offline app and put it online in a way that Just Works. We'll continue to refine this experience in the next few chapters.

Comparing and contrasting with the design principles

We added a lot of new features in this chapter. Syncing lists with an online database, support for user accounts, multiple lists per user, and sharing lists between user accounts. Let's put our phone into airplane mode and do another quick evaluation against the principles with the same pass/fail scoring as before.

Give me uninterrupted access to the content I care about.

This is a pass. Besides maps, the entire app works as expected while offline. Thanks to PouchDB, even when we can't talk to the online database, our offline database works just fine.

Content is mutable. Don't let my online/offline status change that.

This is a pass. The content in our app is perfectly mutable; again, thanks to PouchDB.

Error messages should not leave me guessing or unnecessarily worried.

This is a fail. Even though we're not supporting offline maps in our app, we should provide a better error message. Additionally, there are a bunch of errors that only appear in the developer console and not in the app itself, so sometimes it appears that nothing happened when in reality, we just ran into an error. We'll improve error messaging soon.

Don't let me start something I can't finish.

This is a pass. You can still save changes to the to-do items while offline.

An app should never contradict itself. If a conflict exists, be honest about it.

This is a pass. However, when conflicts do exist, we should show an error message and let you resolve the problem. That's not something that we do right now.

When the laws of physics prevail, choose breadth over depth when caching.

This is a fail. Caching is still pretty dumb. By default, PouchDB caches everything, which works for now, but it isn't scalable.

Empty states should tell me what to do next in a delightful way.

This is a fail. We still don't provide much guidance when you're filling out a to-do item for the first time. Let's address this in the next chapter.

Don't make me remember what I was doing last. Remember for me.

This is fail. We haven't yet provided a way to keep the context sticky. We'll address this problem soon.

Degrade gracefully as the network degrades. Don't be jarring.

This is a fail. When the network goes offline, we're completely silent about it. This isn't jarring right away but we aren't setting clear expectations either.

Never purge the cache unless I demand it.

This is a pass. When you sign out, we purge the cache for you.

Where does it need to improve?

Thanks to PouchDB, we improved from 3/10 to 5/10 in this round, even with the added complexity of being online. As we continue to refine the experience, we can only improve. Most of the current low-hanging fruit is around messaging, which we'll address soon.

Summary

In this chapter, we put our app online. Thanks to PouchDB, we didn't have to sacrifice any of our offline resiliency and got a lot of nice online functionality for not too much extra effort. If you don't have a user account, you can still get value from the app, and if you do have an account, your lists will stay backed up and can be easily shared with others.

In the next chapter, we'll make the app more communicative about errors and set better expectations when switching between the online and offline modes. This will build trust and help the users to not wonder what's going on during the synchronization process.

5
Be Honest about What's Happening

In the last chapter, we put our fledgling app online. Currently, it's a bit tone-deaf. We don't communicate anything that's happening under the hood so there's a lot of *is it working? I really hope it's working*. We also don't attempt to compensate for bad network connectivity. In this chapter, we'll address both of these concerns.

A lot of what differentiates a bad offline experience from a good offline experience is the way an app communicates with people. With two functionally-equivalent apps, the one that comes across as most transparent and honest will win every time. This is particularly important in a first impression scenario. Over time, you may learn to accommodate an app's quirks and foibles, but if people don't stick around after their first bad experience, you've lost.

Fortunately, PouchDB gives us a lot of information about what's happening. We've ignored these events thus far, but in this chapter, we'll be able to pass this information on to people in an unobtrusive way, adding a layer of trust that's currently missing.

Exposing the system state to the user

The first step towards being transparent is to tell people about what's going on in the background. We know when synchronization starts, pauses, and ends. Let's put this data to good use by creating an indicator that shows the current status of synchronization. When errors occur, let's show them too.

Creating an online/offline indicator

Let's override these PouchDB events that we've ignored up until now. When we enabled live synchronization, we set up a bunch of empty handlers that listen for different synchronization events. The first thing that we should do is modify our views in order to show off this data regardless of where the user is in the app.

Adding a global toolbar

First, we need a toolbar that is always visible. This toolbar can't be tied to any individual view, so we'll attach it to the `Main` view instead. To do this, edit `todo-app/app/view/Main.js` and replace the contents of `config` with the following code:

```
layout: 'vbox',
items: [
  {
    xtype: 'panel',
    layout: 'card',
    itemId: 'todo-main-panel',
    items: { xtype: 'todo-lists' },
    flex: 1
  },
  {
    xtype: 'toolbar',
    docked: 'bottom'
  }
]
```

This nests the `card` view in `panel` with `toolbar` docked to the `bottom` of this panel. We'll display any global messages here.

Next, edit the `Main` controller and change `refs` accordingly. Delete the `mainPanel` ref and add the following instead:

```
main: 'todo-main',
mainPanel: '#todo-main-panel',
```

Adding indicator logic

With the view in place, we need to wire up the controller to display messages in the toolbar. The first thing that we'll implement is a basic message that tells us whether the app is online or offline. In the `startSyncing` method, edit the call to `store.localDB.sync()`. Immediately after the `live` and `retry` attributes, add the following code:

```
back_off_function: function (delay) {
  me.online = false;
  me.setIndicator("offline :-(");
  return 1000;
}
```

Whenever the app goes offline, this method will trigger once every second, setting `me.online` to `false`. This variable will be used elsewhere in the controller to indicate whether the app is currently online or not. This method also displays a message in the global toolbar to this effect.

If this is where `me.online` is set to `false`, we obviously need to set it to `true` elsewhere; otherwise, the app will never show as being online. In the `.on('paused')` clause, add the following code:

```
me.online = true;
me.setIndicator("online :-)");
```

Now, after each sync operation (or when the app is first loaded), we'll set the variable to true and display a message indicating that the app is online and operating normally.

Finally, let's create the `setIndicator` method and define the `online` variable. Add the following code after the `startSyncing` method:

```
online: null,
message: null,
setIndicator: function(message) {
  var me = this;

  if (me.message != message) {
    me.getMain().down('toolbar[docked=bottom]').setTitle(message);
    me.message = message;
  }
}
```

This is all pretty straightforward. After we display the message, we set the `me.message` variable to the current message. As our `back_off_function` method is called each second while the app is offline, this avoids redundant component queries.

Commit your changes and then start your web app. Once the app is loaded and you've signed in, you should see a screen similar to the following:

Online app

Displaying error messages

In addition to the event handlers, we have a catch-all method for any errors that take place. Right now, we just dump this to the browser log but this isn't normally something that people see. Instead, we can modify the status indicator in the last section to display these errors when they occur but in an unobtrusive way.

Building trust through up front communication

There are two forms of communication: **passive communication**, which is making people aware of what's going on and **active communication**, which means responding to events in a clear, proactive way. We made use of passive communication in the first section. Our online/offline indicator is always present, letting us know when we're online or offline. However, when people click save or the lists (re)load, there's no feedback from the app. We can do better.

The easiest way to address this is to add a spinner and/or message when people add a new list, collaborator, or item. As each of these actions involves a round-trip to Cloudant, a reasonably expensive and (comparatively speaking) slow event, we should let people know that this is happening and when it finishes successfully (or fails).

This is actually trivial to do using our existing framework. Simply add the following code to the `.on('active')` clause of the `store.localDB.sync()` and call in the `startSyncing` method:

```
me.setIndicator("Syncing…");
```

That's it! Now, whenever you make a change, the app displays a syncing indicator until the operation is complete (at which point, it will revert to the online message).

Predicting the future

Up until now, we've assumed a uniform view of the world. Internet connection, when present, is always strong and reliable. If the app is online, we assume that it can synchronize with little or no trouble. Obviously, this is not always the case. The first step to embracing this is to make the app aware of dead zones in the real world.

When a phone goes from good Internet connectivity to bad or no Internet connectivity, make a note of this for future reference. When the phone nears this location again, use passive communication to warn people of impending network troubles.

Writing a predictive algorithm

We'll use location services for this. Our algorithm works as follows:

- Is our location accurate within 50 meters? If not, exit.

- Are there any points in our database? If not, save the current position (and online/offline state) and exit.

- Are we more than ten meters away from the closest point in our database? If so, save the current position and online state.

- Are we within ten meters of the closest point in our database? If so, update the online state of this point.

- Do we have a velocity over one meter per second? If not, exit.

- Calculate where people will be in ten seconds based on their current velocity. If the closest point to this location is offline, post a warning.

Let's implement this step by step. First, edit the `Main` controller and add a method named `predictBandwidth` under the `init` method:

```
predictBandwidth: function() {
  var me = this;
  me.geo = Ext.create('Ext.util.Geolocation', {
    listeners: {
      locationupdate: function(geo) {
      }
    }
  }
},
```

This method creates a `Geolocation` instance, which triggers the `locationupdate` event every few seconds with the user's current position. As this is done only once, let's add a reference to the method at the end of our `init()` method when the application starts:

```
me.predictBandwidth();
```

Now, define the `geo` variable, next to the `syncHandler` variable that we defined in the last chapter:

```
geo: null,
```

Finally, let's switch back to the `predictBandwidth` method and flesh out the algorithm. Put each of the following code snippets into the `locationupdate` event handler in the given order.

Is the location accurate enough?

The first step is to ensure that our location is accurate enough. For our purpose, we want to have an accuracy within 50 meters with 90% confidence:

```
if (!geo._accuracy || geo._accuracy > 50)
  return;
```

We choose 50 meters for a few reasons. Our algorithm creates a map of points, showing whether there's online/offline connectivity at each point. Wi-Fi is commonly understood to have a 46-meter range while indoors, so our 50-meter threshold accounts for this.

Creating a seed location

Initially, our database is empty. As our algorithm expects at least one point for this to work, we need to populate it in case none exist:

```
var store = Ext.getStore('Position');
if (!store.getCount()) {
  store.add({
    latitude: geo._latitude,
    longitude: geo._longitude,
    offline: me.offline
  });
  return;
}
```

You'll notice that we use a store called `Position`. This store does not exist yet. We'll do this in a bit.

Adding additional points

At this point, we have at least one point in our database. The next step is to add additional points, given that they are at least ten meters apart:

```
var records = store.getData().all,
  closest = null,
  closestDistance = null;
records.forEach(function(e) {
  var distance = Math.sqrt(Math.pow((e.data.latitude -
    geo._latitude), 2) + Math.pow((e.data.longitude -
      geo._longitude), 2));
  if (!closest || distance < closestDistance) {
    closest = e.data;
    closestDistance = distance;
```

```
        return;
      }
  });
  if (closestDistance > 10) {
    store.add({
      latitude: geo._longitude,
      longitude: geo._latitude,
      offline: me.offline
    });
  }
```

One obvious question is how much space this will take. Over time, the store will grow, without bounds. Though we could add some kind of purge algorithm, how much space are we talking about here?

Each point consists of two 8-byte floats and a 4-byte Boolean. Thus, each time a point is added to the store, it grows by 20 bytes. For 20 megabytes (one million points) with points separated ten meters apart in a rectangular layout, this is enough to cover an area of 10,000 square meters (247 acres).

Additionally, consider the rate at which new points are created. In a worst-case scenario (when driving along a road, for example), new points are created each time the locationupdate event handler is called, once every three seconds. Creating one million points at this rate would take 35 days of continuous use of the app.

There's much we could do to optimize this, but for our purposes, this is good enough. I'll list some of the ways you could further improve this if you're interested in developing this algorithm further.

Updating the existing points

So, if we create new points when we're further than 10 meters away from the closest existing point, what about when we're less than 10 meters away? As the online/offline status is not static, let's update the closest point with our current online/offline state:

```
if (closestDistance < 10 && closest.online != me.online) {
  var record = store.findRecord('id', closest.id);
  record.set('online', me.online);
  store.sync();
}
```

Note that the points themselves remain fixed. They don't drift in space. This is to prevent the map from being skewed. For example, if I have ten points in a line that are ten meters apart, I could conceivably drag them away from their origin by updating them in this way.

Ensuring that the velocity is significant enough

This is the final check before we project whether people will go offline soon. As GPS coordinates tend to drift, we'll minimize false positives by requiring people to be moving at one meter per second or greater in order to trigger the warning. This is a casual walking pace.

```
if (!geo._speed || geo._speed < 1)
  return;
```

Predicting future connectivity

Now, we can detect where people will be in ten seconds. If they're using the app, this will give them enough time to pause and finish what they're doing before they go out of range.

```
// Determine where user will be in 10 seconds
var R = 6367444.7, // Earth's radius in meters
  distance = geo._speed * 10, // Current velocity times 10 (seconds)
  dx = distance * (Math.sin(geo._heading * Math.PI / 180)),
  dy = distance * (-Math.cos(geo._heading * Math.PI / 180)),
  dlng = dx / (R * Math.cos(geo._latitude)),
  dlat = dy / R
  newLng = geo._longitude + dlng,
 newLat = geo._latitude + dlat;

// What is the closest point to that location? (nearest neighbor
  search)
closest = null;
closestDistance = null;
records.forEach(function(e) {
  var distance = Math.sqrt(Math.pow((e.data.latitude - newLat), 2)
    + Math.pow((e.data.longitude - newLng), 2));
  if (!closest || distance < closestDistance) {
    closest = e.data;
    closestDistance = distance;
    return;
  }
```

```
  });

  // Was the closest point offline? Post a warning...
  if (!closest.online) {
    this.setIndicator("Going offline soon. :-/");
  }
```

There are three stages here. First, we calculate their future latitude and longitude based on their current position, velocity, and the radius of the Earth. Second, we check our database to see which point is closest to this future location. Third, if this closest point is offline, we warn people that they risk losing connectivity.

Setting our online/offline state

You may have noticed a couple of references to me.online. We need to set this variable to let the algorithm know our online/offline state. Edit the `Main` controller and make the following changes to the `startSyncing` method.

First, set this variable to `false` whenever the `back_off_function` handler is called. Add the following code to this handler:

```
  me.online = false;
```

Next, set this variable to `true` whenever the synchronization is completed successfully. Add the following code to the paused clause of the `localDB.sync()` call:

```
  me.online = true;
```

Finally, define the variable on the `Main` controller itself. Add the following code immediately after the `syncStarted` variable definition:

```
  online: null,
```

Creating the position store

The last thing that we need to do is create a store for our location data. Create a new file named `todo-app/app/store/Position.js` and add the following:

```
  Ext.define('TodoApp.store.Position', {
    extend: 'Ext.data.Store',
    config: {
      model: 'TodoApp.model.Position',
      autoSync: true,
      autoLoad: true
    }
  });
```

Now, let's create the corresponding model for this store. Create `todo-app/app/`
`model/Position.js` and add the following:

```
Ext.define('TodoApp.model.Position', {
  extend: 'Ext.data.Model',
  requires: [
    'Ext.data.identifier.Uuid',
    'Ext.data.proxy.LocalStorage'
  ],
  config: {
    identifier: {
      type: 'uuid'
    },
    fields: [
      'id',
      'latitude',
      'longitude',
      'online'
    ],
    proxy: {
      type: 'localstorage',
      id: 'todoapp-positions'
    }
  }
});
```

Last, edit the `Main` controller and add references to these new classes to the models
and stores arrays. You're done! The unique thing about this store is that it's not
synced to the cloud. One reason is that different devices have different capabilities
to go online. Your phone has Wi-Fi and a cellular network, whereas your laptop
probably only has Wi-Fi. Thus, it doesn't make sense to share.

The downside is that if you switch phones or laptops, you have to rebuild the
database from scratch. That's not an awesome experience. In the future, we could
modify the algorithm to record other details with each location, such as the device
that was online and its capabilities. This is outside of the scope of this book.

Commit your changes and start your app. Walk around for a bit to gather location
data. Notice how the app learns and warns you when you're about to go out of range.

Letting users provide direction

Sometimes, we can't handle these scenarios automatically. We need to capture people's attention and ask them for input. This should be done sparingly and only after we've made a reasonable attempt to handle flaky network conditions ourselves.

Letting the users know when encountering turbulence

For the purposes of this book, we'll define turbulence as any situation where people are actively writing new information to the app and we haven't been able to successfully persist this for the past 30 seconds.

In this scenario, we should bump our notification from passive to active and let people know that we've tried and failed to save their changes. Still, we should keep these notifications unobtrusive and not interrupt anything that they're currently doing.

The simplest way to do this is to display updates at regular intervals in order to keep the user apprised of the current state of any synchronization operation. We'll provide updates at 10 seconds, 30 seconds, 1 minute, and 10 minutes. Similar to the `predictBandwidth` method, we'll create a function that checks the state of the synchronization at regular intervals and updates our global messaging accordingly.

Displaying messages at regular intervals

Edit the `Main` controller and create a new method named `checkForTurbulence`. Put the following code in this method:

```
checkForTurbulence: function() {
  var me = this;

  return setInterval(function() {
    // Are we online and syncing?
    if (!me.online || !me.syncStarted ||
      me.message.indexOf('offline soon') !== -1)
      return;

    var duration = (new Date()).getTime() - me.syncStarted;
    if (duration > 10 * 60 * 1000) {
      me.setIndicator("Are you on GPRS?");
    }
    else if (duration > 60 * 1000) {
      me.setIndicator("Find faster Internet?");
    }
```

```
      else if (duration > 30 * 1000) {
        me.setIndicator("Still working...");
      }
      else if (duration > 10 * 1000 &&
          me.oneTimeMessage === false) {
        me.setIndicator("Please be patient...");
      }
    }, 1000);
  },
```

First, this method uses `setInterval` to check every second whether the app is online and syncing. If so, it calculates the `duration` since the sync operation started and displays an appropriate message.

Now, at the end of the `init` method, add a reference to this method to initialize it:

```
me.checkForTurbulence();
```

Creating a timestamp when the sync starts

The entire algorithm hinges on when the sync operation started. Let's update the `syncStarted` variable whenever a sync operation starts. First, add the variable definition immediately before the `startSyncing` method:

```
syncStarted: null,
```

Now, edit the `startSyncing` method and set this variable to the current time when a sync operation starts. Add the following code to the active clause:

```
me.syncStarted = (new Date()).getTime();
```

Finally, when the sync operation is complete, you'll need to reset this variable. Add the following code to the active clause:

```
me.syncStarted = null;
```

That's it. Commit your changes. If you're using Chrome's device mode, set the network speed to GPRS or regular 2G. Now do something bandwidth-expensive, such as attaching an image to a to-do item. After the sync operation starts, note the messages that appear as the sync continues.

Getting user guidance

If we haven't managed to save anything for 30 minutes, we'll interrupt people briefly to tell them this. Alternatively, you could give people options such as *don't remind me again* if they know they'll have Internet connectivity eventually and don't want to be bothered or *remind me in 30 minutes* if they want to remain apprised of the situation.

Note that these options are possibly overkill for a simple to-do app. In a mission-critical app, with heavy, ongoing communication by multiple users, you'll want to take a more prominent role in re-establishing communication. For our to-do app, this kind of active solicitation is unnecessary so we'll just display a simple modal.

Adding modal to the app

Edit the `checkForTurbulence` method and add the following condition at the top of the `if` tree:

```
if (duration > 30 * 60 * 1000 && me.oneTimeMessage === false) {
  Ext.Msg.show({
    title: "Fasten your seatbelt",
    message: "We've been trying to sync for 30 minutes, but no
      dice. Your data is safe, but you'll want to connect to
      better Internet eventually.",
    buttons: Ext.MessageBox.OK
  });
  me.oneTimeMessage = true;
}
```

Unlike the other `if` statements, this one displays a modal with an **OK** button to dismiss it. We only want to display this message once per sync operation, so there's a `oneTimeMessage` variable to ensure this. Let's initialize this variable immediately before the `checkForTurbulence` method:

```
oneTimeMessage: false,
```

Showing the modal once per sync

As with the `syncStarted` variable, we need to reset the `oneTimeMessage` variable every time a sync operation is complete. Add the following code to the paused clause in the `startSyncing` method:

```
me.oneTimeMessage = false;
```

Other approaches are to show this modal once per session, once per user account, or provide a way for people to turn it off if they don't want to be reminded. The correct approach depends on the importance of being online. For a to-do app, a modal probably isn't necessary at all; while for online multiplayer games, you probably want to nag people more heavily or disable functionality until the connection is restored.

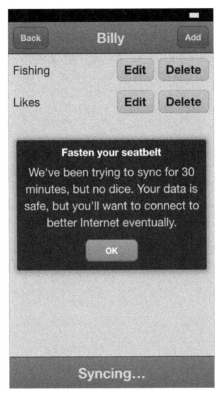

30 minute warning

Staying functional in bad network conditions

Beyond good, two-way communication with people, there are algorithmic things that we can do to compensate for bad Internet connectivity. This boils down to prioritization. In any nontrivial app, there is a large amount of data that we can cache. However, when network bandwidth is limited, we need to decide which data is most important and send this first.

In our to-do app, there are three types of data: images, maps, and text. Of these, text is low-bandwidth, maps are medium-bandwidth, and images are high-bandwidth. In addition, text is the highest priority and maps and images are the lowest priorities.

Thus, we should transmit text first. As maps and images are of equal priority but images are substantially larger, we should transmit maps second and images third. That's the relative priority of each format, but how do we decide which lists/items to cache?

One approach is to keep track of the most commonly viewed lists and cache these first. Another approach is to determine which list is currently being viewed and cache all of the items in this list before any other. A third approach is to cache each data format successively in the order mentioned. These approaches are not mutually exclusive so we'll implement all of them.

First, we'll need to change the way that data is structured in our Cloudant database in order to allow for data retrieval that is as granular as possible. To do this, we'll split our current lists database into four: one for the list metadata, one for the text, one for the map coordinates, and one for the images. In the app itself, we'll sync with each of these databases independently. We could get as granular as the following diagram:

	Current list	Popular Lists	Other Lists
Sync priority	Metadata		
		Metadata	
			Metadata
	Text		
	Maps		
		Text	
		Maps	
	Images		
			Text
			Maps
		Images	
			Images

Order of synchronization

However, for simplicity, we won't differentiate between different types of lists, only the types of data stored in each list: **Metadata**, **Text**, **Maps**, and **Images**. Figuring out how to prioritize the data retrieval by different types of lists is left as an exercise for the reader.

This is how each document is structured. The metadata documents are named with the ID of the list and consist of the following fields:

- The list name
- The owner
- The collaborators
- Last viewed

The text documents are named with the ID of the list, followed by their own ID, and consist of the following fields:

- The item ID
- The description

The map documents are named with the ID of the associated text document, followed by their own ID, and consist of the following fields:

- The item ID
- The latitude
- The longitude

The image documents are named with the ID of the associated text document, followed by their own ID, and consist of the following fields:

- The item ID
- Media

The relationship between these documents looks as follows:

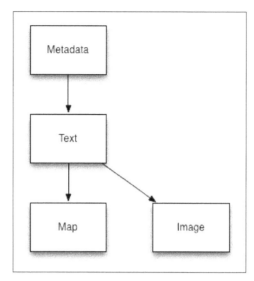

Document synchronization priority

Creating new databases and users

Open your Cloudant dashboard and create four new databases: metadata, text, maps, and images. Take the API users that you created for the lists database and share each of the other databases with them, granting permissions to the reader, writer, and replicator.

The metadata database is what ties everything together. Currently, the app live-syncs with the lists database. Changes are downloaded automatically. However, as we now have five databases with different priority levels, we don't want them to sync willy-nilly. Now, only the metadata database will live-sync and trigger synchronization of the other databases manually (and in the correct order), depending on what needs updating.

Refactoring the Main controller

Now, it's time to implement the code that talks to these new databases. First, because our synchronization code has grown quite large, let's refactor it into its own controller. Create `todo-app/app/controller/Sync.js` and put in the following code:

```
Ext.define('TodoApp.controller.Sync', {
    extend: 'Ext.app.Controller',
```

```
    config: {
    }
}
```

This controller needs access to some of the `views`, `models`, and `stores` from our `Main` controller. We'll add these to the `config` object. First, add the following `views`:

```
views: [
  'TodoApp.view.SignIn'
],
```

Next, add a few `models`:

```
models: [
  'Item',
  'List',
  'User',
  'Position'
],
```

Finally, add the `stores` for these `models`:

```
stores: [
  'Item',
  'List',
  'User',
  'Position'
],
```

Similarly, this controller needs access to some of `refs` and events used by the `Main` controller. Add these to the `config` object as well:

```
refs: {
  main: 'todo-main',
  listsPanel: {
    selector: 'todo-lists',
    xtype: 'todo-lists',
    autoCreate: true
  },
  signInForm: 'todo-sign-in formpanel'
},
```

Now, add a couple of events to handle the sign in/sign out actions:

```
control: {
  'todo-lists button[action=signout]': {
    tap: 'signOut'
```

```
    },
    'todo-sign-in button[action=submit]': {
      tap: 'signIn'
    }
  }
```

This is it for the `config` object. Let's pause and do a little cleanup in the `Main` controller, removing references that are no longer needed by the controller. In the `models` and `stores` configurations, remove the references to `Item`. In the `control` configuration, remove the references to `'todo-lists button[action=signout]'` and `'todo-sign-in button[action=submit]'`.

The next part of the refactoring is to move methods and instance variables that relate to the synchronization. Take the following variables and move them from the `Main` controller to our new `Sync` controller:

- `geo`
- `syncHandler`
- `oneTimeMessage`
- `online`
- `message`

Now, do the same with the following methods:

- `init`
- `predictBandwidth`
- `checkForTurbulence`
- `signIn`
- `signOut`
- `connect`
- `disconnect`
- `startSyncing`
- `setIndicator`

The only method that needs a change is `signIn`. As we still need access to our view-switching logic, we'll need to be explicit about where this function resides. Edit the `signIn` method and replace the last line with the following:

```
this.getApplication().getController('Main').showListsView();
```

Finally, we need to make the application aware of our new controller. Add `Sync` to the controller's array in `todo-app/app.js`. Commit your changes. Start the app and verify that the behavior hasn't changed.

Updating the Item store

The `Item` store needs a lot of work in order to accommodate the new CouchDB databases that we've created. Previously, this store let the `List` store manage it. Now, it will talk directly to Cloudant. When we're done, this store will look very similar to the `List` store.

Defining new databases

Previously, the `Item` store had only a pointer to `currentListStore` and `currentListRecord`. Remove these and replace them with the following database definitions to be used by the methods that we will define next:

```
remoteMetaDB: null,
remoteTextDB: null,
remoteMapsDB: null,
remoteImagesDB: null,
localMetaDB: null,
localTextDB: null,
localMapsDB: null,
localImagesDB: null,
currentListId: null,
```

Don't worry about the fact that they're all `null`. We'll instantiate them in the `Sync` controller later.

Retrieving the data from the databases

First, we need a way to retrieve all the items of a given type for the current list. The following method does exactly this. You provide a store to read and a callback method for any documents that are read. Let's define this method now. Open the `Item` store and add the following method:

```
doWithDocs: function(store, func) {
  var me = this;

  store.allDocs({
    include_docs: true,
    attachments: true,
    startKey: me.currentListId + "_",
```

```
        endKey: me.currentListId + "_\uffff"
      }, function (error, result) {
        func(store, result.rows.map(function(e) {
          return e.doc;
        }));
      });
    },
```

The reason that we return only the items from the current list is efficiency. The app would still work properly, but we'd be passing far more data back and forth than actually needed. At a large scale, the performance would fall apart entirely.

Flagging the databases for sync

As metadata is the only database that live-syncs, we need a way to trigger the synchronization of the other databases manually. The key is to have a method that updates metadata and triggers the synchronization logic when the live-sync occurs.

In the `Item` store, add the following method:

```
flagStoreForSync: function(store) {
  var me = this;
  me.localMetaDB.get(store, function(error, doc) {
    if (!doc) {
      me.localMetaDB.put({
        '_id': store
      });
    } else {
      me.localMetaDB.put(doc);
    }
  });
}
```

All this method does is it updates a document in the metadata database (or creates it if it doesn't exist). This bumps the _rev attribute of this document and triggers a live-sync, which is handled in the `Sync` controller. The rest of the synchronization logic lives in the controller.

Piecing items together from multiple databases

First, add the following event handler to the `listeners` configuration:

```
load: 'onLoad',
```

The onLoad method will do the work of concatenating the text, maps, and images databases into a single Item model, creating new models as needed. Create this method now:

```
onLoad: function(store, records, successful, operation) {
},
```

In this method, add the following code snippets in the order that they appear. First, let's define a closure that retrieves all the images for the current list and attaches them to the appropriate records in the Item store:

```
var me = this;
var updateImages = function(docsArray) {
  me.doWithDocs(store.localImagesDB, function(pouchdb, docs) {
    for (var i = 0; i < docs.length; ++i) {
      docsArray = docsArray.map(function(e) {
        if (e._id == docs[i]._id) {
          e.media = docs[i].media
        }
        return e;
      });
    }
    store.setData(docsArray);
  });
};
```

Now, let's add another closure that does the same thing for the map coordinates. Notice how this closure calls updateImages() when it finishes. As we don't want race conditions, each database is queried in a sequence:

```
var updateMaps = function(docsArray) {
  me.doWithDocs(store.localMapsDB, function(pouchdb, docs) {
    for (var i = 0; i < docs.length; ++i) {
      docsArray = docsArray.map(function(e) {
        if (e._id == docs[i]._id) {
          e.latitude = docs[i].latitude;
          e.longitude = docs[i].longitude;
        }
        return e;
      });
    }
    updateImages(docsArray);
  });
};
```

Next, let's define the last closure in this method, which retrieves the text for all the items in the current list and attaches them to the appropriate records in the `Item` store. In addition, it creates a new `Image` model if the ID doesn't exist and calls `updateMaps()` when done:

```
var setData = function(pouchdb, docs) {
  var docstoadd = [];
  for (var i = 0; i < records.length; ++i) {
    var data = records[i].getData();
    if (docs.every(function(l) { return l._id != data._id })) {
      var model = new TodoApp.model.Item({
        _id: data._id.replace(/.*_/, me.currentListId + "_"),
        list: store.currentListId,
        description: data.description,
        media: data.media,
        latitude: data.latitude,
        longitude: data.longitude
      });
      docstoadd.push(model);
    }
  }
  if (docstoadd.length > 0) {
    docs = docs.concat(docstoadd);
  }
  updateMaps(docs);
};
```

Finally, let's kick off this functionality with a call to `doWithDocs()`:

```
me.doWithDocs(store.localTextDB, setData);
```

Splitting items apart

If `onLoad` joins bits from multiple databases in a coherent set of `Item` models, `onAddRecords` does the opposite. When an item is added, it splits it into text, coordinates, and images, then adds these pieces to the relevant PouchDB database. Remove the existing method with this name and redefine it as follows:

```
onAddRecords: function(store, records) {
}
```

As before, add each of the following code snippets to this method in the given order. First, define a closure that splits the description from each to-do item:

```
var me = this;
var addText = function(pouchdb, lists) {
  var toadd = [];
  for (var i = 0; i < records.length; ++i) {
    var data = records[i].getData();
    if (data.description && lists.every(function(l)
      { return l._id != data._id })) {
      toadd.push({
        '_id': data._id,
        'list': me.currentListId,
        'description': data.description
      });
    }
  }
  if (toadd.length > 0) {
    lists = lists.concat(toadd);
    pouchdb.bulkDocs(toadd, function() {
      me.flagStoreForSync('text');
    });
  }
};
```

Nothing here should be too surprising, but notice the `flagStoreForSync()` method near the end. This method updates the metadata database, which triggers a live-sync and causes the given store (and only that store) to be updated. This will become important later when we need to orchestrate the priority of the various stores.

Now, add a similar closure that splits the coordinates from each to-do item:

```
var addMaps = function(pouchdb, lists) {
  var toadd = [];
  for (var i = 0; i < records.length; ++i) {
    var data = records[i].getData();
    if (data.latitude && data.longitude &&
      lists.every(function(l) { return l._id != data._id })) {
      toadd.push({
        '_id': data._id,
        'list': me.currentListId,
        'latitude': data.latitude,
        'longitude': data.longitude
      });
    }
  }
```

```
    }
    if (toadd.length > 0) {
      lists = lists.concat(toadd);
      pouchdb.bulkDocs(toadd, function() {
        me.flagStoreForSync('maps');
      });
    }
  };
```

Last, define a closure that does the same thing for a to-do item's image:

```
var addImages = function(pouchdb, lists) {
  var toadd = [];
  for (var i = 0; i < records.length; ++i) {
    var data = records[i].getData();
    if (data.media && data.media !== "" && lists.every
      (function(l) { return l._id != data._id })) {
      toadd.push({
        '_id': data._id,
        'list': me.currentListId,
        'media': data.media
      });
    }
  }
  if (toadd.length > 0) {
    lists = lists.concat(toadd);
    pouchdb.bulkDocs(toadd, function() {
      me.flagStoreForSync('images');
    });
  }
};
```

You'll notice that these methods do not create documents needlessly. If a to-do item does not have an image defined, the image will not be created. This saves bandwidth and improves the performance by doing fewer syncs with the remote databases.

Finally, kick off this method with calls to doWithDocs():

```
this.doWithDocs(store.localTextDB, addText);
this.doWithDocs(store.localMapsDB, addMaps);
this.doWithDocs(store.localImagesDB, addImages);
```

Unlike `onLoad()`, we don't need to worry about race conditions. A single store is updating multiple databases and not the other way around, so each closure may be executed in parallel. One downside is that if the Internet connection is poor, it would be better to execute one request at a time. To do this, you could chain the methods together (as `onLoad` does) and flag the metadata store for all three data types at once.

Removing the individual data items

When an `Item` model is deleted, we need to delete the corresponding data in each database. There's already an `onRemoveRecords` method in the `Item` store. Replace it with the following:

```
onRemoveRecords: function(store, records, indices) {
  var me = this;
  var func = function(pouchdb, lists) {
    for (var i = 0; i < records.length; ++i) {
      lists = lists.filter(function(e) { return e._id ==
        records[i].getData()._id; });
    }
    for (var i = 0; i < lists.length; ++i) {
      lists[i]._deleted = true;
    }
    if (lists.length > 0) {
      pouchdb.bulkDocs(lists, function() {
        if (pouchdb === store.localTextDB) {
          me.flagStoreForSync('text');
        } else if (pouchdb === store.localMapsDB) {
          me.flagStoreForSync('maps');
        } else if (pouchdb === store.localImagesDB) {
          me.flagStoreForSync('images');
        }
      });
    }
  };

  me.doWithDocs(store.localTextDB, func);
  me.doWithDocs(store.localMapsDB, func);
  me.doWithDocs(store.localImagesDB, func);
},
```

You should recognize the patterns from `onLoad()` and `onAddRecords()`. We define a closure that finds all the matching records, sets the `_deleted` attribute to `true`, passes them to the appropriate database (with `bulkDocs`), and then flags the store to be synchronized.

Updating the individual data items

You're probably tired of this by now but we have one more case to cover: updating the databases when an `Item` model is updated. Replace the contents of the existing `onUpdateRecord` method with the following, one snippet at a time. First, verify that there's something to update:

```
var me = this;
if (modifiedFieldNames.length == 0) {
  // No changes, don't bother updating the list
  return;
}
var data = record.getData();
```

If there's a description and it differs from the existing description, update the database:

```
if (modifiedValues['description']) {
  store.localTextDB.get(data['_id'], function(error, doc) {
    if (!doc) {
      doc = {
        '_id': data['_id'],
        'list': me.currentListId
      }
    }
    if (doc.description != data.description) {
      doc.description = data.description;
      store.localTextDB.put(doc, function() {
        store.flagStoreForSync('text');
      });
    }
  });
}
```

Similarly, if there's an image and it differs from the current image, update this database as well:

```
if (modifiedValues['media'] !== null) {
  store.localImagesDB.get(data['_id'], function(error, doc) {
    if (!doc) {
      doc = {
        '_id': data['_id'],
        'list': me.currentListId
      }
    }
    if (!doc.media) {
      doc.media = "";
```

```
    }
    if (doc.media != data.media) {
      doc.media = data.media;
      store.localImagesDB.put(doc, function() {
        store.flagStoreForSync('images');
      });
    }
  });
}
```

Finally, do the same for the map coordinates. We don't bother to check whether the coordinates exist. If they don't, we clear the coordinates instead of updating them:

```
store.localMapsDB.get(data['_id'], function(error, doc) {
  if (!doc) {
    doc = {
      '_id': data['_id'],
      'list': me.currentListId
    }
  }
  if (doc.latitude != data.latitude || doc.longitude !=
    data.longitude) {
    doc.latitude = data.latitude;
    doc.longitude = data.longitude;
    store.localMapsDB.put(doc, function() {
      store.flagStoreForSync('maps');
    });
  }
});
```

Updating the list store

The List store does not change as drastically as the Item store, but it still needs a few modifications to work properly. We'll wire up the existing event handlers to work with the flagStoreForSync() method, which works almost identically to the method defined in the Item store.

Creating a pointer to the metadata database

Open the List store. Immediately below the remoteDB and localDB variables, add a reference to the metadata database:

```
localMetaDB: null,
```

This is the same database as that used by the `Item` store. The `flagStoreForSync()` method will use it extensively.

Flagging the databases when they change

In each of the event handlers (with the exception of `onLoad`), we need to let the `Sync` controller know when a specific database needs to sync. In each call to `bulkDocs()` or `put()`, add a second argument with the following code:

```
function() {
  me.flagStoreForSync();
}
```

This will trigger a sync of the `lists` database immediately after any documents are written to it. Now, define the `flagStoreForSync()` method:

```
flagStoreForSync: function() {
  var me = this;
  me.localMetaDB.get('lists', function(error, doc) {
    if (!doc) {
      me.localMetaDB.put({
        '_id': 'lists'
      });
    }
  });
}
```

Updating models and views

To support the changes in the store and reflect the fact that the `Item` store is tied closely to its PouchDB counterparts, make the following changes to the `Item` model:

- Add the `idProperty` attribute to the `config` object:
  `idProperty: '_id'`
- Change the id field to `_id` to match the equivalent PouchDB field
- Add a `_rev` field to match the equivalent PouchDB document revision field
- Add a list field to keep track of which list owns a particular to-do item

Now, let's edit a few of the views to match the changes that you made to the model. First, edit `todo-app/app/view/item/DataItem.js` and change the id field to `_id`.

Next, edit `todo-app/app/view/item/Edit.js` and change the id field to `_id`. Add the following two fields as well:

```
{
  xtype: 'hiddenfield',
  name: '_rev'
},
{
  xtype: 'hiddenfield',
  name: 'list'
},
```

Next, edit `todo-app/app/view/item/New.js` and add the following field:

```
{
  xtype: 'hiddenfield',
  name: 'list'
},
```

Finally, edit `todo-app/app/view/list/List.js` and add the following method to initialize the `Item` store:

```
initialize: function() {
 // Autoload appears to be broken for dataviews
  Ext.getStore('Item').load();

  this.callParent();
}
```

Wiring up the Sync controller

The last thing that we need to do is to make the `Sync` controller aware of all the changes that we just made. One part of this is renaming the variables to match the new naming convention, so be sure to pay close attention. Open the `Sync` controller and let's get started.

Initializing the databases

In the `Sync` controller, replace the `init` method with the following:

```
init: function() {
  var me = this,
    listStore = Ext.getStore('List'),
    itemStore = Ext.getStore('Item'),
```

```
record = Ext.getStore('User').first(),
data;

listStore.localDB = new PouchDB('lists');
listStore.localMetaDB = itemStore.localMetaDB =
  new PouchDB('metadata');
itemStore.localTextDB = new PouchDB('text');
itemStore.localMapsDB = new PouchDB('maps');
itemStore.localImagesDB = new PouchDB('images');

if (record) {
  data = record.getData();
  listStore.username = data.username;
  itemStore.username = data.username;
  me.connect(data.username, data.password);
}

me.predictBandwidth();

me.checkForTurbulence();
},
```

As before, this method is responsible for initializing the PouchDB databases and kicking off various sync operations.

Tweaking the connection logic

Now, edit the `connect` method. Both the `Item` and `List` stores need to know the credentials of the current user. We don't have a central place for these at the moment, so both of the stores get a copy. Add the following lines immediately before the `syncHandler` check:

```
itemStore.username = username;
itemStore.password = password;
```

Now, edit the `disconnect` method. As the `Sync` controller reads the `username` and `password` from the `Item` store, we need to redirect the reset logic in this method. Initialize a variable named `itemStore` with the `Item` store and then replace the `listStore.username` and `listStore.password` lines with the following:

```
itemStore.username = 'nobody';
itemStore.password = null;
```

Changing how the syncing works

Finally, edit the `startSyncing` method. This is where we connect our local PouchDB instances with the remote Cloudant databases, so add the following lines after `listStore.remoteDB`:

```
itemStore.remoteTextDB = new PouchDB('https://' +
  itemStore.username + ':' + itemStore.password +
  '@djsauble.cloudant.com/text');
itemStore.remoteMapsDB = new PouchDB('https://' +
  itemStore.username + ':' + itemStore.password +
  '@djsauble.cloudant.com/maps');
itemStore.remoteImagesDB = new PouchDB('https://' +
  itemStore.username + ':' + itemStore.password +
  '@djsauble.cloudant.com/images');
itemStore.remoteMetaDB = new PouchDB('https://' +
  itemStore.username + ':' + itemStore.password +
  '@djsauble.cloudant.com/metadata');
```

Next, as live-sync occurs with the metadata database instead of the lists database, point the sync to the `localMetaDB` database:

```
me.metaSyncHandler =
  itemStore.localMetaDB.sync(itemStore.remoteMetaDB, {
```

Next, in the `complete` handler of the `syncHandler` variable, reinitialize all of the other PouchDB databases. Add the following lines after the `listStore.localDB` line:

```
itemStore.localTextDB = new PouchDB('text');
itemStore.localMapsDB = new PouchDB('maps');
itemStore.localImagesDB = new PouchDB('images');
itemStore.localMetaDB = new PouchDB('metadata');
```

Finally, replace the `listStore.load()` line in the change handler with the following:

```
me.calculateSync(change.change.docs);
```

This method is where the magic happens. When a change is detected, this method is responsible for determining which databases were affected, syncing these databases, and loading the stores. Let's create this method now:

```
calculateSync: function(metadata) {
}
```

Add the following code snippets to this method in the given order. First, let's define some variables:

```
var me = this,
    itemStore = Ext.getStore('Item'),
    listStore = Ext.getStore('List'),
    syncLists = false,
    syncText = false,
    syncMaps = false,
    syncImages = false;
```

These variables give us access to the stores and PouchDB instances and help us determine which of these to sync. By default, we assume that nothing must be synced.

Next, we iterate through each document. Based on the ID of each document, we flip the sync flag for the related store. Once we've iterated through all of the documents, we know exactly which PouchDB instances to sync:

```
for (var i = 0; i < metadata.length; ++i) {
  var store = metadata[i]._id;
  if (store === "lists") {
    console.log("Sync lists");
    syncLists = true;
  } else if (store === "text") {
    console.log("Sync text");
    syncText = true;
  } else if (store === "maps") {
    console.log("Sync maps");
    syncMaps = true;
  } else if (store === "images") {
    console.log("Sync images");
    syncMaps = true;
  }
}
```

Once done, all that's left to do is the sync operation itself. After the syncing, the related store is reloaded, which ensures that the changes are reflected in the UI:

```
if (syncLists) {
  listStore.localDB.sync(listStore.remoteDB, function() {
    listStore.load();
  });
}
if (syncText) {
  itemStore.localTextDB.sync(itemStore.remoteTextDB, function() {
    itemStore.load();
```

```
    });
  }
  if (syncMaps) {
    itemStore.localMapsDB.sync(itemStore.remoteMapsDB, function() {
      itemStore.load();
    });
  }
  if (syncImages) {
    itemStore.localImagesDB.sync(itemStore.remoteImagesDB,
      function() {
      itemStore.load();
    });
  }
```

Note that these syncs are done in parallel and not prioritized as we eventually want. We'll fix this in the next section.

Prioritizing the synchronization

All that's left to do is prioritize the order in which our databases are synced. When only one database changes, the behavior remains unchanged. It's only when multiple databases are synced that we want to ensure the highest valued data is retrieved first. Delete the last block of code that you wrote in the previous section and replace it with the following:

```
    me.doSync(syncLists, syncText, syncMaps, syncImages);
```

This method recursively iterates through each argument, syncs the database if true, and then moves to the next. It ensures that lists are loaded first, followed by text, then maps, and finally, images. Let's implement this method now:

```
    doSync: function(syncLists, syncText, syncMaps, syncImages) {
      var me = this,
        itemStore = Ext.getStore('Item'),
        listStore = Ext.getStore('List');

      if (syncLists) {
        listStore.localDB.sync(listStore.remoteDB, function() {
          listStore.load(function() {
            me.doSync(false, syncText, syncMaps, syncImages)
          });
        });
      } else if (syncText) {
        itemStore.localTextDB.sync
          (itemStore.remoteTextDB, function() {
```

```
          itemStore.load(function() {
            me.doSync(false, false, syncMaps, syncImages);
          });
        });
      } else if (syncMaps) {
        itemStore.localMapsDB.sync
          (itemStore.remoteMapsDB, function() {
          itemStore.load(function() {
            me.doSync(false, false, false, syncImages);
          });
        });
      } else if (syncImages) {
        itemStore.localImagesDB.sync
          (itemStore.remoteImagesDB, function() {
          itemStore.load();
        });
      }
    },
```

You're done! Commit your changes, start the app, and make some changes; ensure that you add the images, text, and map coordinates. Then, open the app in a second browser. Notice the order in which each data type loads.

Sending the follow-up communication

If, in spite of all of our efforts, we fail to load all the cached data, we should clearly communicate what data is available for use without relying on the spinner of death. This is distinct from the turbulence communication in the last chapter as we're successfully saving some data but just letting the user know what's available and synchronized and what's not.

Comparing and contrasting with the design principles

Let's put our phone into airplane mode and do another quick evaluation against the principles with the same pass/fail scoring as before.

Give me uninterrupted access to the content I care about.

This is a pass. Besides maps, the entire app works as expected while offline. Thanks to PouchDB, even when we can't talk to the online database, our offline database works just fine.

Content is mutable. Don't let my online/offline status change that.

This is a pass. The content in our app is perfectly mutable, again thanks to PouchDB.

Error messages should not leave me guessing or unnecessarily worried.

This is a fail. Even though we're not supporting offline maps in our app, we should provide a better error message. Also, there are a bunch of errors that only appear in the developer console and not in the app itself, so sometimes it appears that nothing happened when in reality, we just ran into an error. We'll improve error messaging in the next chapter.

Don't let me start something I can't finish.

This is a pass. You can still save changes to the to-do items while offline.

An app should never contradict itself. If a conflict exists, be honest about it.

This is a pass. However, when conflicts do exist, we should show an error message and let you resolve the problem. That's not something that we do right now.

When the laws of physics prevail, choose breadth over depth when caching.

This is a fail. Caching is still pretty dumb. By default, PouchDB caches everything, which works for now, but it isn't scalable.

Empty states should tell me what to do next in a delightful way.

This is a fail. We still don't provide much guidance when you're filling out a to-do item for the first time. Let's address this in the next chapter.

Don't make me remember what I was doing last. Remember for me.

This is a fail. We haven't yet provided a way to keep the context sticky. We'll address this problem in the next chapter.

Degrade gracefully as the network degrades. Don't be jarring.

This is a pass. We do a few things to help here. First, we predict when you're about to go offline and warn you about it. Next, we retrieve and display the text first and so in low-bandwidth conditions, you still get value from the app. In addition, when the synchronization takes a long time, we let you know so that you're not caught unawares.

Never purge the cache unless I demand it.

This is a pass. When you sign out, we purge the cache for you.

Where does it need to improve?

Thanks to the improvements in this chapter, we improved from 5/10 to 6/10 in this round. Several of these items are extremely easy to fix, so we can expect the experience to improve drastically in the remaining chapters.

Summary

In this chapter, we exposed the underlying system state, implemented bi-directional communication between the app and people using the app, predicted when the app is about to go offline, made compensation for turbulent network conditions, and split up the database to account for more granular caching behavior.

In the next chapter, we'll address the mother of all synchronization problems. What happens when changes are made to an offline app on multiple devices? Which change wins? How are these conflicts remediated? We'll investigate strategies to deal with this split-brain problem and implement one of them.

6

Be Eventually Consistent

In this chapter, we'll finally address the question of what happens when two users update the same piece of information while offline. This problem, known as split-brain, is common to all distributed systems. When the apps finally get online, which change wins? Do we resolve the conflict automatically or do we let the user decide?

Try this now. Open the app in two separate browsers. Using the mobile development tools, put each app into offline mode. Now, make changes to the same to-do item on each device and then put them online. Notice how the last change wins. This is certainly a solution to the problem, but is it the best solution? Probably not.

We can't know which change is the correct change. Maybe both the users think that their change should win or don't even consider the possibility that their change may fail. We could let the owner win by default, but this doesn't address what happens when two collaborators make changes. We should give users the final say on what happens. When a conflict occurs, show each of them a notification that explains what happened.

To resolve the problem, they have three options. They can either accept the other person's change, overwrite it with their own change (letting the other user(s) know), or create a new to-do item with their change intact, leaving the original to-do item untouched. As we have revision history, a fourth option could be to go back to some previous state. This is outside the scope of this book but an option worth considering.

By the end of the chapter, when split-brain occurs, your users should know exactly what is going on and be able to easily remediate the problem. The to-do app was already consistent but now it will be transparent about the process used to get to this state.

What is a split-brain?

To illustrate the problem, here's a quick diagram:

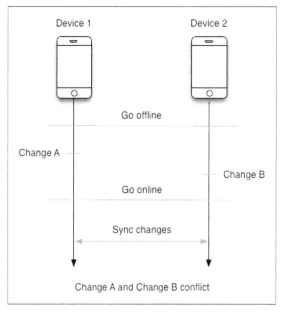

Split-brain

We can mitigate this problem by increasing the granularity of our data storage. In *Chapter 4*, *Getting Online*, the changes were made at the list level. Thus, two people making changes to the same list while offline would always have their changes conflicting. Thanks to the improvements that we made in *Chapter 5*, *Be Honest about What's Happening*, each person would have to make a change to the same description, map coordinate, or image for a conflict to occur.

This is vastly improved. However, when a conflict does occur, how do we resolve it? At its core, there are three basic answers. Either **Change A** wins, **Change B** wins, or both exist in perpetuity. In addition, depending on how much the owners and collaborators of a list trust each other, we could either allow a single person to resolve the conflict or require input from every person with edit privileges. Let's look at both of these scenarios now.

A collective agreement

In this scenario, all the owners and collaborators must agree on the conflict resolution. If a single member rejects the resolution chosen by the others, the conflict will stand. This approach is useful in low-trust environments where we don't want a single user to be able to vandalize a list or exert undue influence:

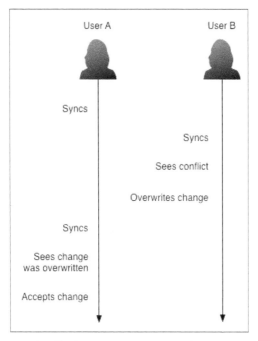

A collective agreement sequence diagram

We can see that the collaborators have the right to overwrite each other's changes. Perhaps **User B** knows something that **User A** does not and wants the list to reflect the current reality. However, overwriting someone else's change is not a silent operation. The person who made the original change can see that their change was overwritten and may choose to accept it, reject it, or save their original change as a separate to-do item.

Trust is a vital ingredient in this approach. Collaborators should know each other and have the ability to police themselves; otherwise, conflict resolution can degrade to everyone continually overwriting each other's changes. Now, let's look at how conflict resolution works at a workflow level:

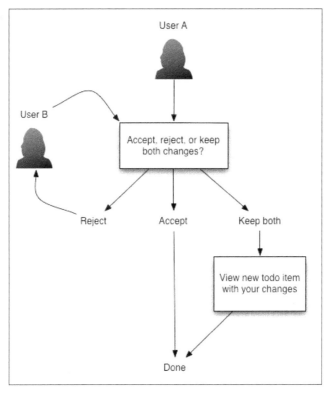

A collective agreement workflow diagram

As long as both the users **Reject** any changes, the loop will continue infinitely. There are two ways to end the loop and resolve the conflict. Either one user must **Accept** the change made by the other or keep both the changes. In the latter case, this creates a new to-do item with the user's changes intact and the original to-do item is left as is.

This is a pretty complicated workflow. In cases where owners and collaborators trust each other to do the right thing, can we simplify things? The answer is yes.

A self-appointed dictator

In this scenario, when a conflict occurs, the first person to address it causes the conflict to disappear. Nobody else will be aware that a conflict ever existed. This is a much lighter approach and one that we will take in our to-do app.

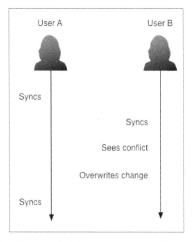

A self-appointed dictator sequence diagram

The nice thing about this is that it's exactly a subset of the collective agreement approach. When you're developing an app, you can start with the lighter approach and upgrade to a more rigorous conflict resolution algorithm as needed. The workflow is simpler as well:

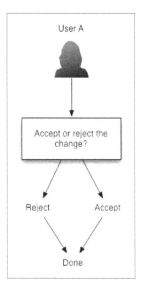

A self-appointed dictator workflow diagram

All that is needed is a way for the user to select the correct change and the app will synchronize this change to all the other users transparently. It accomplishes our goal of making the conflicts visible with a minimum of effort. Before we implement this feature, let's do a bit of refactoring to make these changes easier to implement and to improve performance.

Refactoring the item store

We've made a lot of changes to the `Item` store very quickly and without much attention to the code duplication or sync efficiency. Before we go further, let's refactor the store so that it's easier to change going forward.

Restricting sync to the user's lists

Right now, when we flag each database to be synchronized, we flag it for all the lists—even the lists that you don't have access to. As more people use the app, this will result in an unacceptable hit to performance. Instead, let's restrict sync to the lists that the current user owns.

Open the `Item` store and edit the `flagStoreForSync` method. Replace each usage of `store` with the following code:

```
store + '_' + me.username
```

Now, open the `List` store and edit the `flagStoreForSync` method here as well. Make the same substitution, but with `'list'` instead:

```
'list' + '_' + me.username
```

Edit the `Sync` controller and modify the `calculateSync` method. Add the following two variable definitions to the beginning of the `for` loop:

```
var store = metadata[i]._id.replace(/_.*/, "");
var object = metadata[i]._id.replace(store + "_", "");
```

Here, we extract the `store` name and `username` from the `_id` attribute. Now, add the following condition to each `if` and `else` statement:

```
&& object === itemStore.username
```

This ensures that the `username` matches the current user. If it doesn't, we do nothing (and save ourselves a useless sync operation). If it does, we sync as usual.

Only set the data if needed

Whenever `setData()` is called, the `onAddRecords` event handler is called, resulting in unnecessary processing and synchronization. We should be smarter about this. We should only set the data when PouchDB and our local store disagree. For now, we'll use the count of the store and the `_id` attribute for this.

Open the `Item` store and add the following helper method:

```
setDataIfNeeded: function(store, docs) {
  var recordIds = store.getData().all.map(function(r)
    { return r.getId(); }).sort();
  var docIds = docs.map(function(r) { return r._id; }).sort();

  if (recordIds.length !== docIds.length) {
    store.setData(docs);
    return;
  }

  for (var i = 0; i < recordIds.length; ++i) {
    if (recordIds[i] != docIds[i]) {
      store.setData(docs);
      return;
    }
  }
},
```

This method takes the current list of records in the store, sorts it by `_id`, and then compares it with the list of documents (after sorting these by `_id` as well). If the lists have the same length and the `_id` attributes are the same for all the elements, it assumes that the lists are the same and doesn't call `setData()`.

Obviously, this ignores the fact that `_id` is not enough to guarantee sameness. Later in this chapter, we'll add support for the `_rev` attribute and check for equivalence there as well.

Now, wherever `setData()` is called in the `Item` store, replace it with a reference to `setDataIfNeeded` (except, of course, in `setDataIfNeeded` itself).

Refactoring the event handlers

Each event handler has an enormous amount of duplicated code. This adds complexity to the store and makes it difficult to change the code in the future. Let's refactor this common code into several helper methods. Edit the `Item` store and make the following changes.

The onLoad method

Add a helper method named `loadDocsAttributes` and give it the following code:

```
loadDocsAttributes: function(me, store, docsArray, attributes,
    flag, callback) {
  me.doWithDocs(store, function(pouchdb, docs) {
    for (var i = 0; i < docs.length; ++i) {
      var indexOf = -1;
      for (var j = 0; j < docsArray.length; ++j) {
        if (docsArray[j]._id == docs[i]._id) {
          indexOf = j;
          break;
        }
      }
      if (indexOf === -1) {
        docsArray.push(docs[i]);
      } else {
        docsArray[indexOf][flag + 'rev'] = docs[i]['_rev'];
        Ext.each(attributes, function(attr) {
          docsArray[indexOf][attr] = docs[i][attr];
        });
      }
    }
    callback(docsArray);
  });
},
```

This method does a few things. First, it calls `doWithDocs()` to get a list of data types in the current list from the specified PouchDB database. Next, it checks whether each of these documents already exists in the `Item` store. If not, it adds them. If so, it copies the new attributes to the existing document. Finally, it sends the updated array of documents to the specified callback.

Now that we've abstracted this out, we can call it from `onLoad` for each data type that we support: text, maps, and images. We'll use this pattern for each of the remaining event handlers. Replace `onLoad` with the following code:

```
onLoad: function(store, records, successful, operation) {
  var me = this;
  var updateImages = function(docsArray) {
    me.loadDocsAttributes(me, store.localImagesDB, docsArray,
        ['media'], 'images', function(docs) {
      me.setDataIfNeeded(store, docs);
    });
  };
```

```
    var updateMaps = function(docsArray) {
      me.loadDocsAttributes(me, store.localMapsDB, docsArray,
        ['latitude', 'longitude'], 'maps', function(docs) {
        me.setDataIfNeeded(store, docs);
        updateImages(docs);
      });
    };
    var updateText = function(docsArray) {
      me.loadDocsAttributes(me, store.localTextDB, docsArray,
['description'], 'text', function(docs) {
        me.setDataIfNeeded(store, docs);
        updateMaps(docs);
      });
    };
    updateText(records.map(
      function(r) {
        return r.getData();
      })
    );
  },
```

This is pretty straightforward. We use the two helper methods that we've created so far, loadDocsAttributes and setDataIfNeeded, in order to take the redundancy out of the method. Using the callback, we chain these methods together such that the text database is loaded in the store first, followed by maps, and images last.

The onAddRecords method

Add a helper method named addDocsAttributes and give it the following code:

```
addDocsAttributes: function(me, store, docsArray, attributes,
  flag, callback) {
  me.doWithDocs(store, function(pouchdb, docs) {
    var toadd = [];
    for (var i = 0; i <docsArray.length; ++i) {
      var exists = docs.some(function(d) { return d._id ==
        docsArray[i]._id; });
      if (!exists) {
        var obj = {
          '_id': docsArray[i]._id,
          'list': me.currentListId
        };
        var change = false;
        obj['_rev'] = docsArray[i][flag + 'rev'];
        Ext.each(attributes, function(attr) {
```

```
            if (docsArray[i][attr]) {
              obj[attr] = docsArray[i][attr];
              change = true;
            }
          });
          if (change) {
            toadd.push(obj);
          }
        }
      }
    }
    if (toadd.length > 0) {
      store.bulkDocs(toadd, function() {
        me.flagStoreForSync(flag);
        callback(docsArray);
      })
    } else {
      callback(docsArray);
    }
  });
},
```

Like `loadDocsAttributes`, this method loads all the data items from the specified database for the current list. It then adds them to a temporary list if they don't already exist in the store. If this list has items in it by the end of the algorithm, they are written to PouchDB using `bulkDocs()` and then the app is flagged for a sync operation. Finally, a `callback` method is called with the list of records to be added.

Now, replace the `onAddRecords` method with the following:

```
onAddRecords: function(store, records) {
  var me = this;

  var addImages = function(docsArray) {
    me.addDocsAttributes(me, store.localImagesDB, docsArray,
      ['media'], 'images', Ext.emptyFn);
  };
  var addMaps = function(docsArray) {
    me.addDocsAttributes(me, store.localMapsDB, docsArray,
      ['latitude', 'longitude'], 'maps', function(docs) {
      addImages(docs);
    });
  };
  var addText = function(docsArray) {
    me.addDocsAttributes(me, store.localTextDB, docsArray,
['description'], 'text', function(docs) {
```

```
        addMaps(docs);
      });
    };
    addText(records.map(
      function(r) {
        return r.getData();
      })
    );
  },
```

The differences between `onAddRecords` and `onLoad` are slight. The `text, maps,` and `images` database are modified in the same order and each use the helper that we've defined. We don't change the data in the store as this handler is only called after `setData()`, the purpose being to replicate any new records to PouchDB to sync.

The onRemoveRecords method

Add a helper method named `removeDocsAttributes` and give it the following code:

```
removeDocsAttributes: function(me, store, docsArray, flag,
  callback) {
  me.doWithDocs(store, function(pouchdb, docs) {
    docs = docs.filter(
      function(e) {
        return docsArray.some(
        function(r) {
            return r._id == e._id;
          }
        );
      }
    );
    for (var i = 0; i < docs.length; ++i) {
      docs[i]._deleted = true;
    }
    if (docs.length > 0) {
      pouchdb.bulkDocs(docs, function() {
          me.flagStoreForSync(flag);
          callback(docsArray);
      });
    } else {
      callback(docsArray);
    }
  });
},
```

After the call to doWithDocs(), this method takes all the documents marked for removal, sets their _deleted attribute to true, and writes these changes to PouchDB with bulkDocs. Finally, it triggers the callback method, passing the list of records to be removed.

Next, replace onRemoveRecords with the following code:

```
onRemoveRecords: function(store, records, indices) {
  var me = this;

  var removeImages = function(docsArray) {
    me.removeDocsAttributes(me, store.localImagesDB, docsArray,
      'images', Ext.emptyFn);
  };
  var removeMaps = function(docsArray) {
    me.removeDocsAttributes(me, store.localMapsDB, docsArray,
      'maps', function(docs) {
      removeImages(docs);
    });
  };
  var removeText = function(docsArray) {
    me.removeDocsAttributes(me, store.localTextDB, docsArray,
      'text', function(docs) {
      removeMaps(docs);
    })
  };
  removeText(records.map(
    function(r) {
      return r.getData();
    })
  );
},
```

This method is pretty much identical to onAddRecords. Our helper method once again does the heavy work and leaves this method free to define the higher-level algorithm.

The onUpdateRecord method

Add a helper method named updateDocsAttributes and give it the following code:

```
updateDocsAttributes: function(me, store, data, attributes, flag,
callback) {
  store.get(data['_id'], function(error, doc) {
    if (!doc) {
      doc = {
        '_id': data['_id'],
```

```
          'list': me.currentListId
        }
      }
    var changes = attributes.some(
      function(attr) {
        if (!doc[attr] && !data[attr]) {
          return false;
        }
        return doc[attr] != data[attr];
      }
    );

    if (changes) {
      Ext.each(attributes, function(attr) {
        doc[attr] = data[attr];
      });
      store.put(doc, function() {
        me.flagStoreForSync(flag);
        callback(data);
      });
    } else {
      callback(data);
    }
  });
},
```

This method is unique in that it only deals with a single record at a time and not several. This means we can call `get()` instead of `doWithDocs()`. If the document doesn't already exist in PouchDB, we create it. If there are any attribute changes, we apply them, then save the document, and `flag` the `store` for the synchronization. Finally, we send `record` to the specified `callback`.

Next, replace `onUpdateRecord` with the following code:

```
onUpdateRecord: function(store, record, newIndex, oldIndex,
  modifiedFieldNames, modifiedValues) {
  var me = this;

  var updateImages = function(data) {
    if (modifiedValues['media'] !== null) {
      me.updateDocsAttributes(me, store.localImagesDB, data,
        ['media'], 'images', Ext.emptyFn);
    }
  };
  var updateMaps = function(data) {
```

```
    me.updateDocsAttributes(me, store.localMapsDB, data,
      ['latitude', 'longitude'], 'maps', function(doc) {
      updateImages(doc);
    });
  };
  var updateText = function(data) {
    if (modifiedValues['description']) {
      me.updateDocsAttributes(me, store.localTextDB, data,
        ['description'], 'text', function(doc) {
        updateMaps(doc);
      });
    } else {
      updateMaps(data);
    }
  }
  updateText(record.getData());
},
```

This method is very similar to the previous event handlers, with one difference. The `updateText` and `updateImages` closures check that the given attributes have changes before passing them off to the handler. This is because we don't want to create empty documents in PouchDB that do nothing but hold empty strings.

That's enough refactoring for now. Commit your changes, start the web server, and play with the to-do app. You'll notice that it has become more efficient when changes occur. Fewer sync operations occur and fewer event handlers in the `Item` store are called, resulting in increased performance.

Implementing conflict detection

PouchDB is aware of conflicts under the hood. By default, when a conflict occurs, it chooses an arbitrary winner and returns it on request. This isn't quite the behavior that we want. Instead, when a conflict occurs, we want PouchDB to give us a list of all of the conflicts so that the user can resolve them. The first step is altering our stores to detect and expose the conflicts.

Getting PouchDB to return conflicts

The first step is to get PouchDB to return a list of the conflict revisions. Edit the `doWithDocs` method in the `Item` store and add the following option to the `allDocs` method immediately after the `endKey` option:

```
conflicts: true
```

Now, each document with conflicts will have a `_conflicts` attribute with an array of all the losing conflicts. Once we have this, we can let the user select the revision that they actually want to win and then remove all of the other conflicts.

Attaching conflicts to the item model

Now that PouchDB is returning conflicts, we need to attach them to the `Item` model so that they'll show up in the edit form. First, edit this model and add the following fields:

```
'textconflicts',
'mapsconflicts',
'imagesconflicts',
```

Each attribute will store an array of revs and be accessible from the `hiddenfield` components in the edit form. Where do these attributes come from? The `_conflicts` attribute from the documents in each of our three databases: `text`, `images`, and `maps`. Let's implement this mapping now.

In the `Item` store, edit the `loadDocsAttributes` method and replace the `indexOf === -1` if block with the following code:

```
if (indexOf === -1) {
  var obj = Ext.clone(docs[i]);
  if (obj['_conflicts'].length) {
    obj[flag + 'conflicts'] = docs[i]['_conflicts'];
  }
  docsArray.push(obj);
} else {
  if (docs[i]._conflicts.length) {
    docsArray[indexOf][flag + 'conflicts'] =
      docs[i]['_conflicts'];
  }
  Ext.each(attributes, function(attr) {
    docsArray[indexOf][attr] = docs[i][attr];
  });
}
```

If we're loading a document that doesn't exist in the local store, update the type-specific rev and conflict attributes of this document. If the document does exist, do the same but reference the existing document instead of creating a new one.

Attaching the current revision to the item model

Conflicts are one thing, but we also need to attach the current revision to the `Item` model. This way, when we submit the changes to PouchDB, we're using the revision that we received during the last sync and not the latest revision, which may or may not map to the data that we loaded. This helps PouchDB to flag the conflicts better.

Edit the `Item` model and add the following fields:

```
'textrev',
'mapsrev',
'imagesrev',
```

Now, just as with conflicts, we can multiplex the `_rev` attribute from each document that we get to the appropriate field. However, in contrast, we do this during submit as well and not just on load. Edit the `Item` store and modify the `loadDocsAttributes` method. Change the `if...else` block that you created in the last section and add the following to the `if` part of the block, immediately before the call to `docsArray.push()`:

```
obj[flag + 'rev'] = obj['_rev'];
obj['_rev'] = undefined;
```

Now, add the following similar code to the end of the `else` part of the block:

```
docsArray[indexOf][flag + 'rev'] = docs[i]['_rev'];
```

This is it for the load. Now, let's ensure that our type-specific revision fields are mapped back to `_rev` when the form is submitted. Edit the `updateDocsAttributes` method and add the following to the top of the `if (changes)` block:

```
doc['_rev'] = data[flag + 'rev'];
```

Now, the revisions and conflicts are being mapped from our PouchDB databases to our `Item` model. With this extra information, we can improve the efficiency of the data loads.

Checking the revision before loading the data

Our `setDataIfNeeded` method is responsible for taking the data that we load from PouchDB and setting the store. Previously, if PouchDB and the store had the same set of the `_id` attributes, we didn't update the store. This works when adding/removing the to-do items but does the wrong thing when just updating an existing item as `_id` remains unchanged.

Now, as we have revision as a part of the `Item` model, we can compare this against `_rev` of each document that we load. If the revisions aren't the same, we know our local store is out of date and can update it. Replace `setDataIfNeeded` with the following:

```
setDataIfNeeded: function(store, docs, flag) {
}
```

Note that we added a `flag` argument to the method. This helps us with the mapping from `_rev` to `flag + 'rev'`. Now, add the following code to this method in the given order:

```
var sortFn = function(a, b) {
  return a._id > b._id ? 1 : a._id == b._id ? 0 : -1;
};
```

This method sorts an array of objects by their `_id` attributes. We'll use this to simplify the comparison algorithm when checking whether the contents of the local store are up to date with PouchDB. Now, let's retrieve the data from our local store and sanitize it:

```
var records = store.getData().all.map(
  function(r) {
    return r.getData();
  }
).sort(sortFn);
```

This converts the data to a sorted list of objects with their attributes at a top level. Now, let's do the same to the data that we loaded from PouchDB:

```
docs = docs.filter(function(n) {
  for (var i = 0; i < records.length; ++i) {
    if (n._id == records[i]._id) {
      return true;
    }
  }
  return !!n[flag + 'rev'];
}).sort(sortFn);
```

We exclude any documents that aren't in the local store and don't have a revision as these don't technically exist as documents in PouchDB as of yet. Next, let's compare the arrays and see if they contain the same number of items:

```
if (records.length !== docs.length) {
  store.setData(docs);
  return;
}
```

This is a shortcut when items are added or removed from the list as we don't need to compare the revisions to tell that the data isn't the same. However, if the lengths are the same, we need to ensure that no items have been updated since the last load:

```
for (var i = 0; i < records.length; ++i) {
  if (records[i]._id != docs[i]._id || records[i]._rev !=
    docs[i]._rev) {
    store.setData(docs);
    return;
  }
}
```

At this point, if none of the checks have triggered, we assume that the stores are the same and do nothing. This prevents redundant calls to `setData`, which slow down the app and result in unneeded synchronization between our local and remote PouchDB instances.

That's it for the method. As we added the `flag` attribute to the signature of the method, we need to go through the `Item` store and add this parameter to all the calls of the function. Add `'text'`, `'maps'`, or `'images'`, depending on what type of data is being used.

Implementing the conflict resolution

At this point, PouchDB returns conflicts when they exist and assigns them to the `Item` model for use by our controllers and views. The next step is to expose this information to the users and let them resolve conflicts when they arise.

Adding the supporting methods

The last thing that we need to add to the `Item` store is a couple of supporting methods. As PouchDB returns a list of conflict revisions but not the data attached to each revision, we need to fetch this explicitly. People need to see what data is attached to each revision in order to select the right one.

Edit the `Item` store and add the following method:

```
getAllConflicts: function(me, store, id, revs, docs, callback) {
  if (!revs.length) {
    callback(docs);
    return;
  }
  store.get(id, { rev: revs.shift() }, function(error, doc) {
    docs.push(doc);
```

```
      me.getAllConflicts(me, store, id, revs, docs, callback);
    });
  },
```

Normally, we would use `allDocs` to return a list of documents, but it only works with the `_id` attribute and not `_rev`. Instead, we'll use recursion to retrieve each revision in sequence and send the set of documents to a `callback` method when we're done.

Next, once the user has selected the revision that they want to win, we need a helper to remove all the other revisions and resolve the conflict. This next method does exactly this:

```
  resolveConflicts: function(me, store, id, revs, callback) {
    store.remove(id, revs.shift(), function(error, doc) {
      if (!revs.length) {
        store.get(id, function(error, doc) {
          callback(doc);
        });
        return;
      }
      me.resolveConflicts(me, store, id, revs, callback);
    });
  }
```

As before, we have to remove each document one at a time. This method calls itself recursively until the operation is complete and then sends the winning document to `callback` that we specify. This is still very fast because we're querying our local PouchDB. Once all the conflicts are removed, we'll synchronize PouchDB again.

Adding fields to the edit views

The first order of business is to add the fields related to the revisions and conflicts to our forms. Edit `todo-app/app/view/Edit.js` and add the following fields immediately after the `description` text field:

```
{
  xtype: 'hiddenfield',
  name: 'textrev'
},
{
  xtype: 'hiddenfield',
  name: 'textconflicts'
},
```

Now, edit `todo-app/app/view/Image.js` and do the same with the following fields:

```
{
  xtype: 'hiddenfield',
  name: 'imagesrev'
},
{
  xtype: 'hiddenfield',
  name: 'imagesconflicts'
}
```

Finally, edit `todo-app/app/view/Map.js` and do this once again:

```
{
  xtype: 'hiddenfield',
   name: 'mapsrev'
},
{
  xtype: 'hiddenfield',
  name: 'mapsconflicts'
},
```

Now that these fields exist, we can access them in our controller and provide the conflict resolution logic. First, we need a container to hold the conflict resolution views. Edit `todo-app/app/view/Edit.js` and add the following to the end of the items array of the `description` fieldset:

```
{
  xtype: 'todo-conflict'
}
```

Now, let's define this `xtype`. Create a new file named `todo-app/app/view/Conflict.js` and add the following code:

```
Ext.define('TodoApp.view.item.Conflict', {
  extend: 'Ext.Container',
  alias: 'widget.todo-conflict',

  config: {
    layout: 'hbox',
    items: [
    ]
  }
});
```

This is an empty `hbox` layout that we'll fill with buttons using the controller. Each button will contain the description associated with a particular conflict. When clicked, we'll make the selected revision the winning revision and delete the others.

Adding controller logic

There are two parts to this. First, when we click to edit a to-do item, we need to detect conflicts and populate the view accordingly. Second, when we resolve an existing conflict, we need to update the store and refresh the view with the winning revision.

Let's add the edit logic first. Open the `Main` controller, modify the `editTodoItem` method, and add the following variable definitions after the existing definitions:

```
pouchdb = store.localTextDB,
textPanel = editForm.down('fieldset[title=Description]'),
conflictPanel = textPanel.down('todo-conflict'),
record = store.findRecord('_id', button.getData()),
mediaData = record.get('media'),
textConflicts = record.get('textconflicts'),
mapsConflicts = record.get('mapsconflicts'),
imagesConflicts = record.get('imagesconflicts');
```

Next, before the call to `showEditView()`, add the following code block:

```
if (textConflicts) {
  conflictPanel.removeAll();
  store.getAllConflicts(store, pouchdb, button.getData(),
    textConflicts.concat(record.get('textrev')), [],
      function(docs) {
    Ext.each(docs, function(d) {
      conflictPanel.add({
        xtype: 'button',
        action: 'accept',
        text: d.description,
        data: d._rev
      });
    });
    conflictPanel.setHidden(false);
    textPanel.down('textfield[name=description]').setHidden(true);
    textPanel.setTitle('Description (resolve conflict)');
  });
} else {
```

```
        conflictPanel.setHidden(true);
        textPanel.down('textfield[name=description]').setHidden(false);
        textPanel.setTitle('Description');
    }
```

First, this code checks whether there are text conflicts or not. If so, it retrieves the conflicts using the `getAllConflicts` method that we created earlier. Then, it reveals the conflict panel and adds buttons to it, one per revision, each with the description associated with that revision. It also hides the input field and changes the description of fieldset. This ensures that the conflict is resolved before any additional changes are made.

Next, open the `Sync` controller and add the following handler to the control configuration:

```
    'todo-edit button[action=accept]': {
        tap: 'acceptChange'
    }
```

Now, implement the `acceptChange` handler. Add the following code to the `Sync` controller:

```
    acceptChange: function(button) {
    }
```

Add each of the following code snippets to this method in the given order. First, let's define a few variables:

```
    var store = Ext.getStore('Item'),
    pouchdb = store.localTextDB,
    fieldset = button.up('fieldset'),
    text = fieldset.down('textfield[name=description]'),
    rev = fieldset.down('hiddenfield[name=textrev]'),
    conflicts = fieldset.down('hiddenfield[name=textconflicts]'),
    values = button.up('formpanel').getValues(),
    record = store.findRecord('_id', values._id),
    toremove = record.get('textconflicts').concat
      (record.get('textrev')).filter(function(r) {
      return r != button.getData();
    });
```

Now, let's call the `resolveConflicts` method and pass a callback to it in order to update our view:

```
    store.resolveConflicts(store, pouchdb, values._id, toremove,
      function(doc) {
        fieldset.down('todo-conflict').setHidden(true);
```

```
fieldset.setTitle('Description');
text.setHidden(false);

text.setValue(doc.description);
rev.setValue(doc._rev);
conflicts.setValue(undefined);

record.set('description', doc.description);
record.set('textrev', doc._rev);
record.set('textconflicts', undefined);
record.setDirty();
store.sync();

Ext.getStore('Item').flagStoreForSync('text');
});
```

This callback does the following three things:

- First, it hides the conflict-related UI and reveals the fieldset as it normally appears
- Second, it sets the values of the underlying form fields with the winning revision
- Finally, it sets the values of the underlying record, saves it, and triggers a PouchDB sync

Trying it out

First, commit your changes to Git. Now, start the web app in two browsers. Make sure that a to-do item is defined and then put one of the browsers in the offline mode. Make changes to both the to-do items and save. At this point, the instance of PouchDB on each browser has a different revision of the same to-do item.

Enable the online mode in both the browsers. Wait for PouchDB to sync and notice that one of the changes wins by default and applies to both automatically. This is PouchDB's default behavior. Regardless, a conflict exists. Edit the to-do item in one of the browsers and notice that the conflict resolution UI shows both the conflicts.

Choose the losing conflict. Notice how the conflict resolution UI disappears, the app syncs, and the selected revision appears in both the browsers. Refresh to ensure that the resolution succeeded. Now, repeat the process and choose the default winning conflict instead. The same thing should happen, except that the descriptions will stay as they were.

Conflict resolution

Conflict resolution for maps and images

At this point, conflict resolution is fully functional for the description field in each to-do item. Obviously, this technique can be extended to support maps and images as well. The main difference is the UI used to resolve the conflict. Text is easy as a single button can be used to show each revision and mark it as the winning revision.

Maps and images are harder. The content is rich, so you'd likely need to show a thumbnail of each revision, with the option to see the whole thing (if they differ slightly). If you want to implement this, give it a try. The store will need no modification and the controller will need slight modifications based on the UI that you want to construct. I will leave this as an exercise for the reader.

Comparing and contrasting with the design principles

Let's put our phone into airplane mode and do another quick evaluation against the principles with the same pass/fail scoring as before.

Give me uninterrupted access to the content I care about.

This is a pass. Besides maps, the entire app works as expected while offline. Thanks to PouchDB, even when we can't talk to the online database, our offline database works just fine.

Content is mutable. Don't let my online/offline status change that.

This is a pass. The content in our app is perfectly mutable, again thanks to PouchDB.

Error messages should not leave me guessing or unnecessarily worried.

This is a fail. Even though we're not supporting offline maps in our app, we should provide a better error message. Also, there are a bunch of errors that only appear in the developer console and not in the app itself, so sometimes it appears that nothing happened when in reality we just ran into an error. We'll improve error messaging in the next chapter.

Don't let me start something I can't finish.

This is a pass. You can still save changes to the to-do items while offline.

An app should never contradict itself. If a conflict exists, be honest about it.

This is a pass. Thanks to the improvements in this chapter, when conflicts exist, we let the users know and allow them to select a resolution.

When the laws of physics prevail, choose breadth over depth when caching.

This is a fail. Caching is still pretty dumb. By default, PouchDB caches everything, which works for now, but it isn't scalable. We'll fix this in the next chapter.

Empty states should tell me what to do next in a delightful way.

This is a fail. We still don't provide much guidance when you're filling out a to-do item for the first time. Let's address this in the next chapter.

Don't make me remember what I was doing last. Remember for me.

This is a fail. We haven't yet provided a way to keep the context sticky. We'll address this problem in the next chapter.

Degrade gracefully as the network degrades. Don't be jarring.

This is a pass. We do a few things to help here. First, we predict when you're about to go offline and warn you about it. Next, we retrieve and display the text first, so in low-bandwidth conditions, you still get value from the app. In addition, when the synchronization takes a long time, we let you know so that you're not caught unawares.

Never purge the cache unless I demand it.

This is a pass. When you sign out, we purge the cache for you.

Where does it need to improve?

We still have four areas where the experience falters but the improvements in this chapter buttressed the existing areas and we'll do our best to fix the remaining issues in the next two chapters. We've solved most of the hard problems. Getting to 10/10 is easier than you might think at this point.

Summary

In this chapter, we discussed different ways of dealing with a split-brain: the collective agreement approach where every collaborator must agree on the same resolution and the self-appointed dictator approach where any collaborator can choose a resolution independently of the others. We implemented the second approach for the description field of the to-do items and discussed ways to extend this to the map and image fields as well.

In the next chapter, we'll look at ways to optimize the performance of the app. Up until now, we've stuck with the standard defaults provided by our offline database. Let's investigate where our defaults fall short and how to improve them. We'll also learn how to detect different usage scenarios and tweak the defaults in real time to accommodate these scenarios.

Choosing Intelligent Defaults **7**

In the previous chapter, we addressed the problem of conflicts that occur when we are offline. Thanks to our offline PouchDB instance, we were able to track revisions while offline and reconcile the changes when we are back online. When multiple resolutions exist, we let the user choose which one to keep and dispose of the others. In the future, if we implemented an undo feature, we would see a linear revision history with no branches.

In this chapter, we will look at the low-hanging fruit that we've ignored up until now. Many of these things are trivial. When we load the app while offline, the Google Maps API leaves an ugly error message. When no to-do lists or to-do items exist, the screen is empty instead of providing a hint of what to do next. When the user reopens the app, they probably want to come back to whatever they were viewing last.

These are simple things and we can fix them easily. There are a few other defaults that involve a little more work, but we should fix them as well. First, PouchDB caches everything by default. This was fine when we started, but because storage is limited, we need to set limits and respect these limits. By default, we should cache all the list metadata (otherwise you can't navigate to a to-do list) and cache all the data in the current list, but caching other data is up to the discretion of the app.

This app discretion gives us some rope to work with. A second improvement that we can make is to track the lists that are viewed most frequently. Once our top cache priorities are out of the way, we'll cache the data from the other lists in the appropriate order. Additionally, when we exceed our cache limits, we can remove items from the cache in the reverse order.

Finally, if people want to clear their cache for any reason, we'll give them a straightforward way to do so. To inform this behavior, we could show them how much data is being consumed by the database and disable this functionality when the database is already empty, for example.

Beyond this, we'll talk about different scenarios that people deal with when their cache is being loaded with data. We'll discuss how these scenarios affect the performance of the app and ways in which you might tweak the behavior of your own app to compensate for certain edge cases.

Low-hanging fruit

The first six chapters focused on the harder problems in offline-first web development. There are a couple of reasons for this. First, a lot of apps do the small obvious things right (mostly), so there's not much to learn there. Second, solving the harder problems first means that you're making those changes while the app is simpler to understand and modify.

Small changes, even ultimately positive ones, add complexity to an app, making it harder to make changes later. We're now at a point where we can afford to backtrack and solve these smaller problems.

Error messages

Remember when we implemented support for maps in *Chapter 3, Designing Online Behavior*? As we haven't added support for offline maps, we get the following error message when opening the app while offline. It's ugly and doesn't describe the problem really well. We can do better.

Current error when map is not available

The current error message is defined by Sencha Touch. Let's override this. Create a new file named `todo-app/app/patch/MapsError.js` and add the following code:

```
Ext.define('TodoApp.patch.MapsError', {
  override: 'Ext.Map',

  constructor: function() {
    this.callParent(arguments);

    if (!(window.google || {}).maps) {
      this.setHtml('Must be online to use Google Maps');
    }
  }
});
```

Commit your changes and start the app in offline mode. Open a to-do item with a map and verify that this more sensible error message is displayed:

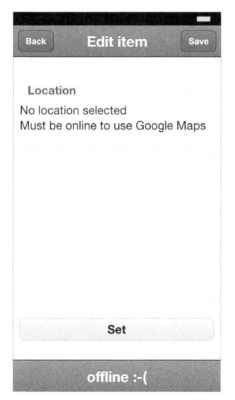

Improved error message when map is not available

Empty states

Next, we need to give our data views an empty state. There are three data views: collaborator.List, list.List, and list.Lists. Sencha Touch provides a configuration option where we can specify a string to be displayed when the data view is empty. Let's do this now.

Edit todo-app/app/view/collaborator/List.js and add the following attribute to the data view:

```
emptyText: '<div style="' +
  'text-align: center;' +
  'height: 100%;' +
  'margin-top: 50%;">' +
  'Add a collaborator to this list' +
  '</div>',
deferEmptyText: true
```

Now, do the same for todo-app/app/view/list/List.js:

```
emptyText: '<div style="' +
  'text-align: center;' +
  'height: 100%;' +
  'margin-top: 50%;">' +
  'Add a todo item to this list' +
  '</div>',
deferEmptyText: true
```

Finally, add an empty string to the data view in todo-app/app/view/list/Lists.js:

```
emptyText: '<div style="' +
  'text-align: center;' +
  'height: 100%;' +
  'margin-top: 50%;">' +
  'Create a todo list and help your brain' +
  '</div>',
deferEmptyText: true
```

Commit your changes and start the app. Delete any existing lists and notice the empty state. Create a list, then click the **Share** and **Edit** buttons. Verify that these empty states show up as well:

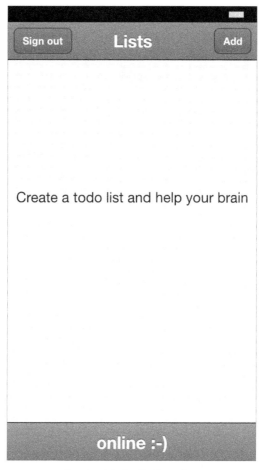

Empty state for the lists view

Restoring the last view

The last bit of low-hanging fruit is to restore the last view that a user was on when the app is restarted. This will help them regain their previous context quickly when the app restarts. To do this, we need two things: a local store to persist this data between the restarts and a way to keep track of the method used to display a given view.

On the extreme end, we could restore any screen the user was on. More realistically though, the user probably only cares about being taken back to the last list or item that they were viewing. Clicking **Edit** again isn't a huge hardship. If we implemented an autosave of the drafts, we could prompt the user to pick up where they left off. For now, we'll support going back to one of the three views: the list of lists, a specific list of the to-do items, and a specific to-do item.

Creating a store and model

First, create a new file named `todo-app/app/model/Restore.js`. In this file, add the following lines:

```
Ext.define('TodoApp.model.Restore', {
  extend: 'Ext.data.Model',
  requires: [
    'Ext.data.identifier.Uuid',
    'Ext.data.proxy.LocalStorage'
  ],
  config: {
    identifier: {
      type: 'uuid'
    },
    fields: [
      'id',
      'currentListId',
      'currentItemId'
    ],
    proxy: {
  type: 'localstorage',
  id: 'todoapp-restore'
    }
  }
});
```

This code defines a basic model backed by local storage, which contains the IDs of the list and/or item to restore. If both of these items are null, no restoration will take place. We'll implement this logic in the controller.

Now, let's create the corresponding store for this model. Create another file named `todo-app/app/store/Restore.js` and add the following code:

```
Ext.define('TodoApp.store.Restore', {
  extend: 'Ext.data.Store',
  config: {
    model: 'TodoApp.model.Restore',
```

```
        autoSync: true,
        autoLoad: true
    }
});
```

Nothing new here. The synchronization and loading is handled automatically, so all we need to do is add a record and the store will handle the rest.

Specifying which pages to restore

Now that our store and model are in place, the next step is to populate these with data when visiting certain pages in the app. On loading, the app will check the store and if the data is present, will load the appropriate pages. Edit the Main controller, then create the following helper method:

```
saveView: function(currentListId, currentItemId) {
    var store = Ext.getStore('Restore');

    store.removeAll();
    store.add({
        'currentListId': currentListId,
        'currentItemId': currentItemId
    });
},
```

This method simply takes currentListId and currentItemId and passes them to the store to be persisted. As we can have only a single page to restore, it clears all the other records from the store. Now, we can add hooks elsewhere in the app that will call this method when the current list or item changes.

Edit the goBack method and append the following code:

```
var xtype = this.getMainPanel().getActiveItem().xtype;
if (xtype === 'todo-list') {
    this.saveView(Ext.getStore('List').currentListId, null);
}
else if (xtype === 'todo-lists') {
    this.saveView(null, null);
}
```

This code checks the xtype of the panel that we've gone back to. If it's a list panel, we tell the app to restore that list on loading. If it's a list of lists, we tell the app to do nothing (as this is the default landing page anyway).

Now, edit the `showView` method and add the same code:

```
var xtype = this.getMainPanel().getActiveItem().xtype;
if (xtype === 'todo-list') {
  this.saveView(Ext.getStore('List').currentListId, null);
}
else if (xtype === 'todo-lists') {
  this.saveView(null, null);
}
```

This covers two cases: going back and going to either the list view or list of lists view. Lastly, we need to save the state when the user edits a to-do item. Edit the `editTodoItem` method and add the following code immediately before the call to `showEditView()` at the end:

```
this.saveView(store.currentListId, itemId);
```

This is it. Now, let's look at restoring the state from this data when the app loads.

Loading the page when the app starts

With the information that we've stored, it can detect where we left off and load the appropriate page automatically the next time the app loads. Edit the `Main` controller and add the store and model that you just created to the `models` and `stores` configuration attributes. Now, find the `editTodoItem` method and replace its arguments with the following:

```
editTodoItem: function(itemId) {
  ...
}
```

This method now receives a to-do item's identifier directly, rather than having to extract it from the button control. The reason we do this is because this method will be called from two places: when a to-do item's **Edit** button is clicked and when the restoration data is loaded on the starting of the app. Now, replace each instance of `button.getData()` in the method with `itemId`.

Next, we need to create the equivalent event handler method, which extracts the identifier from a button's data attribute and passes it to `editTodoItem`. Add the following method to the controller:

```
onEditTodoItemClick: function(button, e, eOpts) {
  this.editTodoItem(button.getData());
},
```

Now, go to the `control` attribute in `config` and replace the call to `editTodoItem` with a call to `onEditTodoItemClick`.

The next step is to do exactly the same thing with the `editList` method. Replace the definition of `editList` with the following code:

```
editList: function(listId) {

    ...
}
```

Replace each instance of `button.getData()` in the method with `listId`. Now, create the corresponding event handler:

```
onEditListClick: function(button, e, eOpts) {
    this.editList(button.getData());
},
```

Finally, in the `control` attribute of the controller's `config`, replace the call to `editList` with `onEditListClick`. This concludes our changes to the `Main` controller. Now, edit the `Sync` controller and add the following attribute immediately before the `doSync` method.

This attribute is very important. It signals whether we've done a restore since the app has been loaded. As restoring the state is a disruptive operation (we don't want people to suddenly be hoisted to a different screen), we want this to happen first thing. Note that `doSync` is called at the end of the `startMetaSyncing` method. As this method is called on `init`, `doSync` is called as well.

Edit the `doSync` method and add the following `else` clause to the end of the `if` statement:

```
} else if (me.restoreOnLoad) {
    var store = Ext.getStore('Restore'),
      record;

    if (store.getCount()) {
      record = store.getAt(0);
      if (record.get('currentListId')) {
        me.getApplication().getController('Main').editList
          (record.get('currentListId'));
      }
      if (record.get('currentItemId')) {
        me.getApplication().getController('Main').editTodoItem
          (record.get('currentItemId'));
      }
    }

    me.restoreOnLoad = false;
}
```

After all of our PouchDB stores have loaded, this clause will be executed if it hasn't already. It loads the list ID and item ID from the store and calls the appropriate methods in the `Main` controller to display the relevant view. However, because `doSync` is a recursive method, we need to make sure that it's called once more after the last PouchDB store loads. Replace the `itemStore.load()` method in the `syncImages` branch of the `if` statement with the following:

```
itemStore.load(function() {
  me.doSync(false, false, false, false);
});
```

That's it! Commit your changes, start the app, navigate to a list or to-do item, and then restart the app. After a couple of seconds, it will load the screen that you viewed previously.

Obviously, this isn't perfect. As the initial sync takes a few seconds, there's a lag in the restoration process. Ideally, we would show a splash page during the initial sync and then show the UI after any screen restoration has taken place. Regardless, this is a good demonstration of the concept.

Cache limits

File space is limited. Databases have a tendency to grow out of bounds, so if you're not careful, you'll end up with more database than you can store. PouchDB is particularly bad in this regard as it stores the revision history of every document. Whenever you make a change, you store both the old document and the new document.

Now, we know that the revision history exists to help us resolve conflicts gracefully in offline scenarios. However, to do this, we only need to save the leaves, the last revision of a document. Is there a graceful way to get rid of older revisions and save space?

Fortunately, the answer is yes. When you create a local PouchDB database, you can set the `auto_compaction` option to make this behavior default whenever a document is written. Edit the `Sync` controller and add the following variable definition to the `init` method:

```
options = { auto_compaction: true },
```

Now, add this variable to each `new PouchDB` statement in the method:

```
listStore.localDB = new PouchDB('lists', options);
listStore.localMetaDB = itemStore.localMetaDB = new
  PouchDB('metadata', options);
itemStore.localTextDB = new PouchDB('text', options);
itemStore.localMapsDB = new PouchDB('maps', options);
itemStore.localImagesDB = new PouchDB('images', options);
```

Do the same thing with the `startMetaSyncing` method. Add an `options` variable to the top of the method and append this variable to each instance of `new PouchDB` in the `complete` event handler at the end of the method. It's worth pointing out that `auto_compaction` has no effect on remote databases, so you'll need to run a remote process to prune these databases if you need to conserve space.

Obviously, this technique doesn't solve the problem that databases grow slowly over time. PouchDB doesn't provide an easy way to limit the amount of space that it consumes. However, individual databases are cheap, so you could give each list its own database and only load databases as needed (or up to some predetermined limit). This would serve the purpose of restricting the amount of space taken up locally by your data.

List frequency

Beyond limits, you need to think about how this space is used. As with limits, PouchDB doesn't come with a lot of control here. You have to build the system yourself. The general pattern holds. Assign one list per database, embed the frequency metadata in a master database (that you always sync), and then choose which databases to sync locally based on this data.

Clearing the cache

Every once in a while, you just want to clear everything in your local cache and start over. Thanks to replication, your data will remain secure; it'll just be wiped from your mobile device. If you sign in again, the data will be synced back to your device. The reason we don't do this when signing out is to speed up the rate at which your device resynchronizes with remote changes; though you can certainly clear it if you want. Clearing the cache is the nuclear option, rarely needed, but important to know how to do.

Google Chrome

To clear the cache in Chrome, do the following:

1. Open Chrome.
2. Open the **History** menu.
3. At the bottom of the menu, click on **Show Full History**.
4. Click on **Clear browsing data...**.
5. Check **Cookies and other site and plug-in data**.
6. Check **Cached images and files**.
7. Click on **Clear browsing data**.

Mozilla Firefox

To clear the cache in Firefox, do the following:

1. Open Firefox.
2. Open the **History** menu.
3. Click on **Clear Recent History...**.
4. Check **Cookies**.
5. Check **Cache**.
6. Click on **OK**.

Apple Safari

To clear the cache in Safari, do the following:

1. Open Safari.
2. Open the **History** menu.
3. At the bottom of the menu, click on **Clear History...**.
4. Click on **Clear History**.

Microsoft Edge

To clear the cache in Edge, do the following:

1. Open Edge.
2. Select **Hub | History**.

3. Click on **Clear all history**.

4. Click on **Clear**.

Usage scenarios

We've gone through a lot of information in the past couple of chapters about intelligent caching. Let's pause and talk about some real-world scenarios that might affect the approach that you take in your own apps. We'll use the scenarios in *Chapter 1, The Pain of Being Offline* and apply what we've learned so far about offline development.

To facilitate the discussion, let's map each scenario to the following three axes:

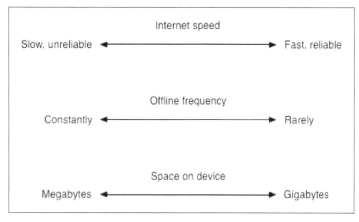

Usage scenario axes

The three dimensions are **Internet speed**, **Offline frequency**, and **Space on device**. The more a scenario falls on the left-hand side of the graph, the more optimization we need to do. If the **Internet speed** is **Slow**, users are going to need caching to be highly prioritized: the most frequently viewed lists and text first and images and revision history last. If they're offline most of the time, the cache becomes sacred. Once gone, it's impossible to get it back. If their device is space-constrained, caching must be highly selective. Frequently viewed items should be cached and items that are never viewed should be purged to keep the data usage slim.

Let's look at some of the scenarios now.

Going on an off-the-grid vacation

Once a year, Carl disappears into the wilderness for a two-week trek along the Pacific Crest Trail in the Western United States. He does this to get away from technology. Sometimes, due to the nature of his job, he has to respond to an important e-mail or take the occasional phone call.

There is no cellular coverage out in the pines. To connect, Carl has to leave the forests behind and hitch a ride to the nearest hotel. After a hot shower and some real food, he cracks open his laptop and endures shoddy hotel Wi-Fi for the evening. It's an inconvenience but a compromise that Carl is willing to accept.

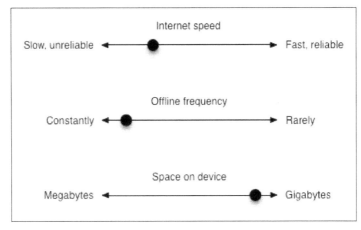

Carl's scenario map

Any apps that Carl uses must have highly-prioritized caching and preserve the contents of the cache at all costs. However, because Carl has a modern smartphone, he can store essentially limitless amounts of data on it.

Living in a third-world country

Marsha owns an old Nokia phone with an i9 keypad. She has access to a slow GPRS network, which often goes down. As modern websites are so bandwidth-intensive, her phone so slow, and online connectivity so tenuous, she turns off images and media when she browses.

In the next town, the mobile network is slightly better, but it's a five-mile walk. When Marsha needs to e-mail photos, she composes draft e-mails and queues them on her phone. Once a week, she treks into town to deliver milk from their small herd of cattle. During this time, she is able to send those e-mails. It's a slow and frustrating experience but she can't do much about it.

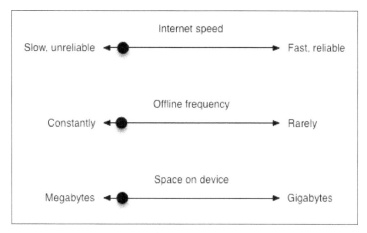

Marsha's scenario map

Any apps that Marsha uses must have highly-prioritized and highly-selective caching, yet cache as much as possible. Due to this, Marsha probably needs an option to trim the cache to a certain size, rather than nuke it entirely, due to the expense of repopulating the cache. Better yet, the app could give her the option to specify the cache size and have the app respect this size automatically.

Commuting on public transportation

Elizabeth takes the bus an hour each way on her daily commute. The bus doesn't have Wi-Fi and encounters several dead zones on the route. She is a dedicated bibliophile, so these routes are a great opportunity to feed her reading addiction. As she reads 1-3 books a week, she often runs out of reading material.

When this happens on the bus, she can't usually do anything about it. Her life is so busy that she doesn't usually have time to think about downloading new books before her commute. Maybe she'll bring a paperback along next time for something to do.

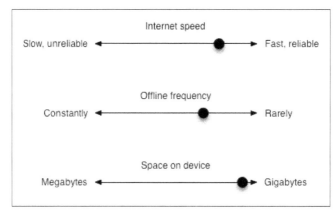

Elizabeth's scenario map

Unlike Carl and Marsha, going offline is more of an annoyance than a serious problem for Elizabeth. She doesn't need her cache to be highly optimized, so long as it stores enough data to keep her occupied in the few instances when the Internet is not available.

Working on a Wi-Fi only device

Shawon is 14. On his last birthday, his parents bought him an iPod Touch. He wanted an iPhone for the coolness factor but, oh well. It's fine. It doesn't have GPS or a mobile plan, but most of the time it's as good as a real iPhone.

Like most kids, he relies on his parents for transportation and accompanies them on their errands. As he has a Wi-Fi only device, he often goes somewhere new, only to have to figure out what the Wi-Fi password is, that is, if the place even has Wi-Fi. There aren't many Wi-Fi networks without passwords these days.

When he can't find the password, he has to stay contented with the games and music that are on his device. However, most of his games require an Internet connection, which seems odd, particularly for games with a single-player mode.

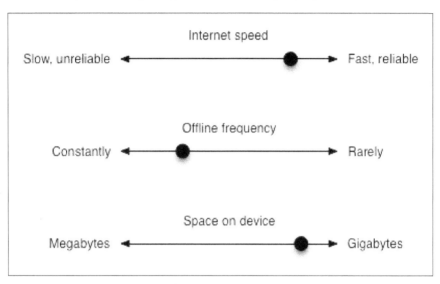

Shawon's scenario map

This scenario is similar to Elizabeth's in that Shawon has a good Internet connection and plenty of space on his device. However, he's much less likely to have an Internet connection because of his Wi-Fi only device. His apps need to have deep caching to give him plenty to do when he's offline.

Sharing files between mobile devices

Francine likes her world where file sharing is so easy and seamless. In 2005, you used USB drives. In 2015, you share files wirelessly with technology and services such as **Dropbox**, **Google Drive**, and **Airdrop**. Unfortunately, when her Internet connection dies, she's transported back to 2005.

Just last week, her editor asked for an update to an outline she wrote. Traveling at the time, she didn't have easy access to Wi-Fi. She made the changes on her laptop but couldn't transfer the file to her phone to be e-mailed. After struggling with this for several minutes, her Airbnb host replied with instructions to connect to the network. Crisis averted.

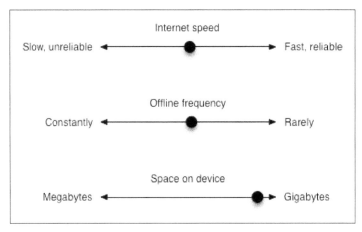

Francine's scenario map

This is a slightly different scenario than the last few ones. Most of the time, people have a pretty good idea of when they'll be without Internet. Business travel is unique in that you're almost never sure of the kind of connection you'll have once you get to your final destination. Francine needs her caching to be deep when an Internet connection isn't present and prioritized to get the most out of a flaky connection.

Streaming high-definition videos from YouTube

Brian streams online videos, a lot. Most of this streaming is over his phone. His mobile provider is kind of horrible, so he has to deal with dodgy Internet connectivity. Most of the time, YouTube is very good about decreasing the quality of the videos to compensate for bad network conditions but it isn't always enough.

He wishes there was a way to get a transcription of the video or just wait for the entire thing to download before playing. YouTube won't download an entire video when it's paused. Unfortunately, these features don't exist. Instead, Brian bookmarks the videos to watch at home.

In this scenario, Brian is rarely offline. Prioritized caching is the only thing that he really needs. As he consumes videos at a faster rate than they can be cached, he wants to be notified when a video is cached and ready to be viewed. Caching video streams is outside the scope of this book.

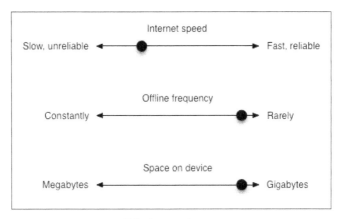

Brian's scenario map

Online shopping with your phone

Jasmine is not an organized person. She likes taking care of things in the moment as they come to mind. As a busy professional, she doesn't have the mental resources to memorize lists and doesn't want to be tied to a to-do app or sheet of paper.

Shopping is one example. Throughout her day, she does a mental inventory as she moves about. When she spots an item in short supply, she grabs her phone and places an order with her Amazon app. When she's connected to the Internet, this works great. When she's not, it frustrates her that she can't take care of things. If only Amazon provided a better add-to-cart experience for offline users.

This use case is very close to the to-do app that we've been developing. Let's see how it maps:

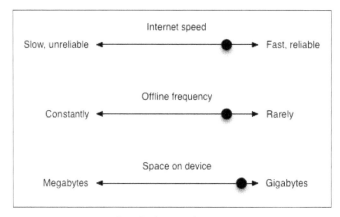

Jasmine's scenario map

Jasmine finds herself offline just enough for it to be annoying. In these scenarios, she has a task that she wants to perform online and can't. Caching is important only to the extent that it captures her intent and allows the app to perform the action when it is back online. When shopping, she might tag the items that she buys frequently, then she might want to go back and place an order for these items. By caching these (and the actions associated with them), she can.

Getting work done at a conference

Conferences are notorious for their abysmal Wi-Fi. Raphael is attending Whole Design along with 1,500 other attendees. It's a three-day, single-track conference. On day two, with four hours to go, he is ready to be done fighting for the Wi-Fi bandwidth. He hasn't gotten much done (besides listen to some great talks). Even his cell phone, normally reliable, doesn't work here at all.

Every couple of hours, the attendees get a 30-minute break. Raphael uses these opportunities to sprint outside the venue. Here, he can get a solid 3G connection. This is enough to check his e-mail, send chats, and generally check in with his world. At least, it should be. His apps behave as though there's a binary switch between no bandwidth at all and enough to stream HD video. A third option would be nice.

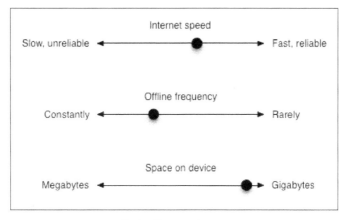

Raphael's scenario map

This scenario is very like the first one. The difference between Raphael and Carl is the interval spent offline. Raphael spends hours while Carl spends weeks. Due to this, prioritized caching is a must but the cache is less sacred as Raphael can refresh it on a fairly regular basis.

Comparing and contrasting with the design principles

Let's put our phone in the airplane mode and do another quick evaluation against the principles with the same pass/fail scoring as before.

Give me uninterrupted access to the content I care about.

This is a pass. Besides maps, the entire app works as expected while offline. Thanks to PouchDB, even when we can't talk to the online database, our offline database works just fine.

Content is mutable. Don't let my online/offline status change that.

This is a pass. The content in our app is perfectly mutable, again thanks to PouchDB.

Error messages should not leave me guessing or unnecessarily worried.

This is a pass. In this chapter, we improved the error message that occurs when maps are not available.

Don't let me start something I can't finish.

This is a pass. You can still save changes to the to-do items while offline.

An app should never contradict itself. If a conflict exists, be honest about it.

This is a pass. Thanks to the improvements in this chapter, when conflicts exist, we let the users know and allow them to select a resolution.

When the laws of physics prevail, choose breadth over depth when caching.

This is a fail. Caching is still pretty dumb. By default, PouchDB caches everything, which works for now, but it isn't scalable. We'll fix this in the next chapter.

Empty states should tell me what to do next in a delightful way.

This is a pass. In this chapter, we added basic empty states to our list views. Users no longer get a blank stare when no data is present.

Don't make me remember what I was doing last. Remember for me.

This is a pass. In this chapter, we implemented stickiness for some of the views namely, the list of lists view, list of to-do items view, and edit a to-do item view.

Degrade gracefully as the network degrades. Don't be jarring.

This is a pass. We do a few things to help here. First, we predict when you're about to go offline and warn you about it. Next, we retrieve and display the text first, so in low-bandwidth conditions, you still get value from the app. In addition, when the synchronization takes a long time, we let you know so that you're not caught unawares.

Never purge the cache unless I demand it.

This is a pass. When you sign out, we purge the cache for you.

Where does it need to improve?

Thanks to the low-hanging fruit that we addressed in this chapter, our score improved significantly as compared with the last chapter. The only area with a fail score is with respect to the caching depth. Instead, we've discussed a variety of scenarios to inform how rigorous your caching algorithm should be and you should use your best judgment when selecting one for your app. Our app scores 9/10 at this point.

Summary

In this chapter, we addressed some low-hanging fruit to improve the score of our to-do app. In addition, we took the scenarios from *Chapter 1, The Pain of Being Offline* and analyzed each against three criteria that affect a user's offline experience: connection speed, offline frequency, and device storage. No one strategy fits every user's needs, but knowing who your users are and what scenarios they experience can help inform the kind of caching that you implement.

In the next chapter, we'll look at ways to keep your data synchronized when you lack an Internet connection. Modern smartphones have a wide variety of network interfaces, including Bluetooth and peer-to-peer Wi-Fi. Let's see how we can keep our PouchDB databases synchronized using these interfaces.

8
Networking While Offline

Up until now, our work has followed a predictable pattern. Present a problem, figure out how to solve it, and implement the solution. This chapter is different. The problem is real, but the technology isn't in place to actually solve the problem.

The basic premise behind each chapter is that the Internet is required to sync your data. You can modify it and create new data offline, but to make it stick, you have to be able to connect to your Cloudant databases.

In this chapter, we will cover the following topics:

- What it means to be offline
- Device support
- Platform-independent libraries
- Synchronization over Wi-Fi
- Making the setup less painful

What it means to be offline

This basic assumption that the Internet will always be there is a faulty assumption. The Internet is best described as a graph of servers, each capable of talking to the others:

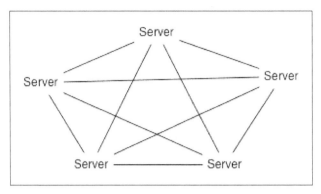

Network topology

A **Server** is anything that talks on the Internet: your laptop, your mobile phone, the machine hosting your Cloudant database, and so on. Offline simply means that a specific server is no longer able to talk to the other servers on the Internet. Thus, offline is a matter of perspective.

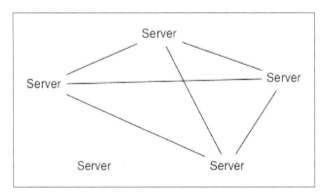

Server offline

Now, this graph is somewhat dishonest. If you're on a mobile device, all the data goes through a wireless router. If this router goes down, all the phones sending their data through this node go offline. This is why, when your Internet goes down, it's common to ask other people if their Internet is down as well.

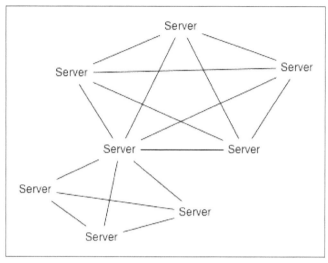

Server hierarchy

If you're the only person affected, your device is probably at fault. If other people are affected at the same time, the fault lies higher in the hierarchy. Sometimes, entire segments of the Internet go down, affecting all the servers that depend on the Internet.

To further complicate the issue, a device may communicate with multiple data sources at the same time. If any of these go down, the apps that depend on them will enter an offline state, but the other apps will remain online. Thus, it can be more useful to consider offline connectivity on an app-by-app basis rather than for the device as a whole.

So how does this relate to our app? In typical usage, there is a set of mobile devices all communicating with each other through our Cloudant instance:

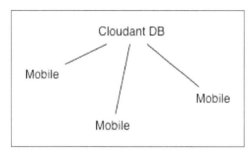

To-do app architecture

Unlike the mesh diagram that we showed earlier, these mobile devices cannot communicate with each other directly. Instead, they write the data to their local PouchDB instances, which synchronize with Cloudant, triggering an update on the other mobile databases.

 Before I explain why things work this way, understand that PouchDB does not mandate this model. In the world of PouchDB, every database is a peer. We synchronize with Cloudant out of convenience, but there's nothing that says that we can't synchronize mobile devices directly, avoiding Cloudant entirely.

I'm not saying that this is necessarily a superior approach. Cloudant is always there. When you have Internet connectivity, you always have something to synchronize with (assuming that Cloudant isn't down). On the other hand, there's no guarantee that any other mobile devices are online to sync with. Also, if you only have one mobile device to your name, Cloudant serves as a backup for your data. You don't have to rely on other devices for backups.

This becomes interesting if you're offline. If you're offline due to being off the grid (a not so uncommon scenario), how do we synchronize your device with another device belonging to you or one of your collaborators?

Today, you can't. At least, not in a platform-independent way, using the facilities provided by modern mobile platforms. Peer-to-peer communication has always been important but very hard to do well. The widespread adoption of the Internet has made peer-to-peer communication less important but correspondingly less considered as well. What we want is a graph as follows:

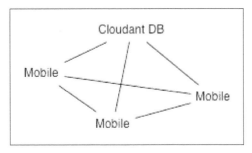

Desired to-do app architecture

Then, when a mobile device goes offline (that is, can no longer talk to Cloudant), it can still synchronize with the other mobile devices in the mesh. In the real world, you often end up with a partitioned network as follows:

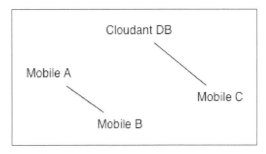

Partitioned network

In this scenario, **Mobile A** and **Mobile B** can still talk to each other. Either they're on the same Wi-Fi network (which has lost Internet connectivity) or are off the grid and relying on some form of mesh network (for example, Bluetooth or peer-to-peer Wi-Fi). They can synchronize their to-do items as they would when online.

In addition, **Mobile C** is still online and able to talk to **Cloudant DB** though it can't talk to **Mobile A** or **Mobile B**. When the connectivity is restored, PouchDB can reconcile any conflicts.

Device support

Unfortunately, this is where we run into problems. Support for Internet-free sharing of data is extremely sparse on mobile platforms. Android supports peer-to-peer Wi-Fi, **Near Field Communication** (**NFC**), Bluetooth, and USB. However, this is specific to Android. If you want access to these features on iOS or Windows Phone, you're out of luck. Let's start by inspecting the support provided by Android, then evaluate a couple of libraries that can be used to emulate this functionality on other platforms.

Open `http://developer.android.com/guide/topics/connectivity/` in your browser and note the support for the following interfaces.

Bluetooth

Bluetooth is a protocol with a typical range of 10 meters. If you're sharing data with someone in the same room, Bluetooth is a good choice. Let's see what the sharing process looks like:

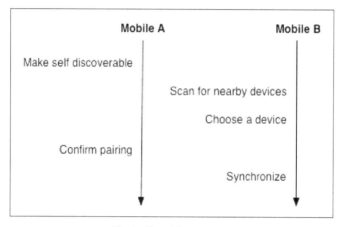

Bluetooth pairing process

There are a few things that we could do to make this process simpler. First, we could turn on discoverability automatically. Second, we could perform periodic scans in the background. Third, we could save pairings, so you only need to pair the first time. Finally, we could make choosing a device and synchronizing the same action (assuming that the devices are already paired). This results in the following optimal workflow:

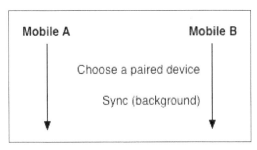

Bluetooth sharing (simplified)

One unintended side effect is that we lose Cloudant's ability to mediate the authentication. We need a way to distribute the authentication so that our offline apps know who has permission to edit which lists. Obviously, we don't want to share these secrets with all the devices that share a copy of our PouchDB database.

PouchDB provides a feature that allows you to store local documents. A local document is never replicated and exists only in the database on which it was created. We can store private information here, such as **Secure Shell** (**SSH**) keys and credentials, without fear that others will get them. The particular encryption method that you use is up to you, but I recommend some form of hashing algorithm.

NFC

NFC is used to describe a variety of protocols, each with a shorter range than Bluetooth—typically 10 cm or less. This is a good distance for two people who are next to one another.

Unlike Bluetooth, there are many incompatible implementations of NFC, so this is probably most useful as a backup to Bluetooth unless you know the hardware specifications of the devices that will be using your app in advance. Additionally, NFC is anywhere from 10 to 100 times slower than Bluetooth, so take this into consideration as well.

NFC has the concept of tags. Tags are small blocks of memory, which can be read by another device without turning on the host device. This is why you can pay for merchandise with your phone without turning it on. Depending on the storage requirements of your app, you could theoretically let another person synchronize with your phone without turning it on. Pretty slick.

Also, as NFC requires such close proximity between devices, this can be a security benefit though you'll probably want to require a passcode or biometric check before permitting the data to be synchronized.

Peer-to-peer Wi-Fi

Peer-to-peer (or ad hoc) Wi-Fi is a technique that allows two devices to share the data directly without going through a router. As with normal Wi-Fi, it has a range of approximately 100 meters, which allows people in the same home to share data. In addition, Wi-Fi has a greater throughput than either Bluetooth or NFC, allowing for quick synchronization.

The process of pairing is very similar to Bluetooth. However, you only need to enable peer-to-peer Wi-Fi; a confirmation of each pairing is not required. In addition, Wi-Fi has the ability for one device to act as a router, allowing other devices to connect and form their own mesh network. This permits a group of people to synchronize their apps simultaneously, something that neither Bluetooth or NFC can do easily.

Platform-independent libraries

Now that we've had an overview of the different protocols available, let's look at a few libraries to help you implement these connections. First, the bad news: there are no platform-independent options for Bluetooth, NFC, or USB. The good news: platform-independent synchronization over Wi-Fi is possible as long as you're willing to manually create an ad hoc network on one device and connect to it from the other device.

USB

From a security and reliability standpoint, it's hard to beat wires. The primary problem with USB is that not all mobile devices include a USB port. However, if a port is available, this is a great way to share data between two devices. It avoids the complicated pairing of wireless protocols, relying only on a hard-wired connection.

You'll still want authenticate, but you don't have to worry about other people eavesdropping on the connection—something that all wireless connections have to worry about (though the shorter the range, the less of a concern this is).

Bluetooth

It's theoretically possible to connect iOS devices together over Bluetooth, but this requires a Made For iPhone certification. For more information about the **MFi** certification, see the following website: `https://developer.apple.com/programs/mfi/`

It is not currently possible (as of Android 5.1.1 Lollipop) to connect iOS and Android devices together, though once a bug is resolved on the Android side, it will be possible to do so using a low energy Bluetooth. The bug report for this issue can be found at `https://code.google.com/p/android/issues/detail?id=58725`

NFC

Currently (as of iOS 8.4), Apple has not exposed NFC support to third-party developers. In contrast, Android support for NFC is quite good, and you can find many Cordova plugins that support this. Unfortunately, as none of these support iOS, you can't depend on NFC as a platform-independent technology. Here's one plugin: `https://github.com/chariotsolutions/phonegap-nfc`

Wi-Fi P2P

Unfortunately, there are no libraries that support this fully as managing Wi-Fi is a system feature that the OS doesn't want apps to be able to override. Android lets you scan for all the visible networks, allowing you to prompt the user to select one. This is only for Android; iOS does not let you scan networks through their APIs.

 Here's one plugin that you can use for Android: `https://github.com/HondaDai/PhoneGap-WifiInfoPlugin`

In iOS 7, Apple introduced a framework for **Multipeer Connectivity**. This allows you to easily expose services using a combination of Wi-Fi networks, peer-to-peer Wi-Fi, and Bluetooth. This is iOS-specific, but if you only care about this platform, you might find it helpful. No Cordova plugins exist for this functionality, but you can see the raw API documents here:

`https://developer.apple.com/library/ios/documentation/MultipeerConnectivity/Reference/MultipeerConnectivityFramework/`

Beyond this, the only way to share data over Wi-Fi in a platform-independent way is to have a device that supports the creation of ad hoc networks and then connect to this network from other devices. As iOS devices don't support the manual creation of ad hoc networks, you'll need to pair them with a device that does so (for example, a laptop).

USB

Apple has famously not provided iOS devices with a USB port. Thus, cross-platform synchronization over USB isn't possible. However, Android has a good support for USB-based data transfer. See the following page for more information:

```
http://developer.android.com/guide/topics/connectivity/usb/
```

Synchronization over Wi-Fi

Now that we've surveyed the landscape, Wi-Fi seems to be our best option for offline synchronization. Let's go through the process of setting up an ad hoc network, synchronizing PouchDB over this connection, and then reconciling any conflicts when we go back online.

Setting up the P2P connection

These instructions assume that you have an OS X laptop. The instructions for Windows are similar. On your laptop, do the following:

1. Click on the Wi-Fi status icon in your status bar:

Wi-Fi status icon

2. Click on **Create Network...**.

3. Name the network Todo App and click on **Create**.

   Notice how the Wi-Fi status icon has changed:

Peer-to-peer Wi-Fi status icon

Now, connect to this network on your iPhone (or other mobile device).

1. Open **Settings**.
2. Choose **Wi-Fi**.
3. Select the **Todo App** network.
4. Click on **Join Anyway**.

You've now created a P2P network between your laptop and mobile device. You're also offline. Next, we'll implement support for offline syncing between these devices.

Assigning an IP address to each device

When you connect to an ad hoc Wi-Fi connection, you have to assign an IP address to each device, manually. Traditionally, when you connect to a router, the router has a **Dynamic Host Configuration Protocol (DHCP)** server that assigns you an IP address automatically when you connect. We don't have this. You could use something such as **Bonjour** or **Zeroconf** to make this easier, but setting these services up is outside of the scope of this book.

Assign an IP address to your laptop using the following steps:

1. Open **System Preferences**.
2. Choose **Network**.
3. Select **Wi-Fi**.
4. Click on **Advanced...**.
5. Choose **TCP/IP**.
6. Choose **Manually** from the **Configure IPv4** drop-down menu.
7. Enter **192.168.1.1** in the **IPv4 Address** field.
8. Enter **255.255.255.0** in the **Subnet Mask** field.
9. Click on **OK**.

Assign an IP address to your iOS device using the following steps:

1. Open **Settings**.
2. Choose **Wi-Fi**.
3. Click on the **i** icon next to the **Todo App** entry.
4. Choose the **Static** tab.
5. Enter **192.168.1.2** in the **IP Address** field.
6. Enter **255.255.255.0** in the **Subnet Mask** field.

Testing the connection

Start the to-do app on the laptop. Instead of localhost, use the IP address that you configured for the laptop:

```
$ open 192.168.1.1:1841
```

Now, see whether your mobile device can view the to-do app on this IP address. Open Safari (or the mobile browser of your choice) and navigate to this IP address. You should see the app get loaded normally:

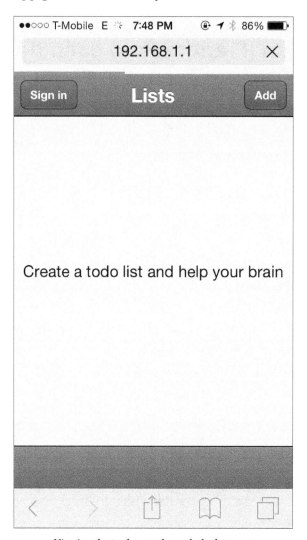

Viewing the to-do app through the browser

Implement support for offline syncing

Making these devices talk to one another isn't enough. CouchDB communicates through a **Representational State Transfer** (**REST**) API. In order to sync across devices, we need to expose this API over at least one of the offline devices that we want to sync. Unfortunately, there's no way to expose an HTTP socket on iOS (and doing so on Android is finicky), so syncing two mobile devices together isn't possible. You'll need a laptop.

1. Open `http://couchdb.apache.org`.

2. Click on **DOWNLOAD**.

3. Choose the appropriate version for your platform and download it.

4. Once the download is completed, run the application. (On OS X, you may need to right-click the file and click **Open**.)

Now, you have an instance of CouchDB running on your local machine. Let's create the databases needed by our app. This setup should exactly mirror what we've configured on our remote Cloudant instance:

1. Click on **Create Database**.

2. Enter `metadata` in the **Database Name** field.

3. Click on **Create**.

4. Repeat these steps for the remaining databases: lists, text, maps, and images.

To keep these databases in sync with our Cloudant instance, you'll want to set up a continuous replication between them:

1. Under the **Tools** menu, click on **Replicator**.

2. Under **Replicate changes from**, choose **Local database**.

3. Choose **metadata** from the drop-down menu.

4. Under **Replicate changes to**, choose **Remote database**.

5. Paste the URL to the remote metadata database. Substitute your username and password: `https://username:password@username.cloudant.com/metadata`

6. Check the **Continuous** box.

7. Click on **Replicate**.

Wait for a few seconds. If you see the following **ok** event displayed, the replication is working correctly:

Replication success

Repeat these steps for the remaining databases: lists, text, maps, and images. This sets up a unidirectional replication from your local CouchDB to Cloudant. To make the replication bidirectional, repeat all of the steps once more but replicate the changes from the remote database to the corresponding local database.

When you're done, click on **Status** under the **Tools** menu; you should see a set of replication events similar to the following:

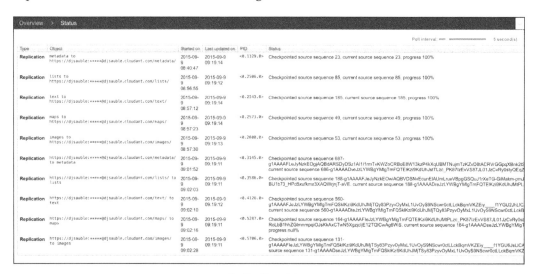

Replication status

Now that your local CouchDB is up and running correctly, the last thing that you need to do is switch the URLs in your to-do app so that they point at the local instance instead of Cloudant. This is obviously a contrived example. We'll talk about ways to make this setup less painful in the next section.

Edit `todo-app/app/controller/Sync.js` and replace the remote DB initialization lines with the following (You may want to comment the existing lines as this is just a temporary change.):

```
listStore.remoteDB = new PouchDB('http://192.168.1.1:5984/lists');
itemStore.remoteTextDB = new
  PouchDB('http://192.168.1.1:5984/text');
itemStore.remoteMapsDB = new
  PouchDB('http://192.168.1.1:5984/maps');
itemStore.remoteImagesDB = new
  PouchDB('http://192.168.1.1:5984/images');
itemStore.remoteMetaDB = new
  PouchDB('http://192.168.1.1:5984/metadata');
```

Now, rebuild the to-do app and reload it on your laptop and mobile device. Notice how the devices continue to synchronize while offline.

Making the setup less painful

Technically, we've enabled offline synchronization between devices, but our approach is littered with caveats:

- Offline syncing only works over Wi-Fi
- Requires a laptop to perform synchronization
- Involves a manual setup of a separate CouchDB database, a peer-to-peer Wi-Fi connection, and static IP addresses
- Cannot sync two mobile devices together (unless syncing through a laptop)
- Have to edit the source code to switch to offline mode

A lot of these reasons are due to technical constraints. Let's go through each point and understand what the ideal solution would be, whether possible today or not. There's a huge opportunity for mobile device manufacturers to improve the current state of offline synchronization in a platform-independent way.

Multiprotocol sync

What if any of the wireless protocols on your phone could be used to sync with other phones? Apps could choose the right one based on the security, energy, and performance requirements of the application, and synchronization would just work with a pairing workflow identical across all the wireless protocols:

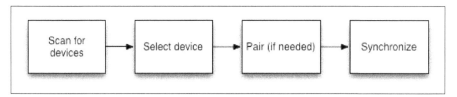

Platform-independent sync workflow

In this way, users remain unaware of the protocol used. All they care about is their ability to share data. Whether it's accomplished over Wi-Fi, Bluetooth, NFC, or pigeon carrier is irrelevant.

Furthermore, this workflow is simple. If the scanning and synchronization occur automatically in the background, the user only needs to select a device and pair with it. The best way to encourage behavior is to remove friction from the equation. This solution does exactly this.

Services on a mobile

Phones and tablets have replaced laptops and desktops in popularity. These devices run on less powerful operating systems and aren't capable of hosting the types of services that a server, desktop, or laptop can. However, what if they could?

Earlier in the chapter, we showed the way the to-do app architecture is designed today:

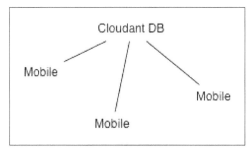

To-do app architecture

A big reason for this is that mobile devices have historically not hosted services the way a laptop, desktop, or server does now. Mobile devices are designed to be as energy-efficient as possible. Hosting an always-on web server runs contrary to this ideal. This is why, in our preceding example, we required a laptop in order to initiate offline synchronization.

Any solution must take the limitations of the mobile into account. The synchronization should be done only when needed, as quickly and efficiently as possible, and in the background. This last bit is especially important. If we allow devices to synchronize only when people are using them, we add a lot more friction to the experience.

Here are a few examples:

User A creates a to-do list and then turns off the phone.

User B opens the app, but because User A isn't using their phone, the app fails to sync.

Now User A and User B are interacting with different sets of data.

 The old-school approach of *I've turned my phone on, now you turn your phone on and let's sync* puts too much effort on people. It's vital that mobile operating systems provide a way for third-party apps to synchronize over wireless protocols without the phones being physically turned on.

This might result in additional restrictions. For example, when phones are off, only low-energy Bluetooth or NFC may be used, whereas when phones are on, Wi-Fi may be used. This is a performance compromise but an acceptable one. Data may synchronize more slowly, but because people aren't using their phones, they won't notice. In this scenario, reserving battery life is more important than synchronizing data as quickly as possible.

The end result is what we want: mobile phones capable of forming mesh networks while offline and synchronizing the data among themselves without being turned on. The technology is here; we just need Apple, Google, and Microsoft to pay attention.

Zero configuration networking

It's not acceptable to expect people to know how to configure an ad hoc network. Devices should just know how to connect to each other in a secure and simple way. There are many competing protocols that accomplish this but very few that do so in a platform-independent way.

Distributed networking is an approach where every device is a peer, capable of self-organizing in a coherent network. The only important thing is the connection between devices. As long as you can trace a line from one device to another through a series of intermediate devices, the network will function as any other. In this scenario, being offline simply means that there's no way to trace a connection to the device that you want to talk to.

> The downside is that there are few platform-independent libraries that implement decentralized networking. DNS and DHCP are by far the most popular approaches, and there's little interest in replacing them with a competing network stack.

In iOS 7, Apple released the Multipeer Connectivity framework. This framework doesn't quite implement the mesh network that we've described but does provide a basic ability to discover other devices nearby and connect to these devices, all while offline. Obviously, the industry still has a long way to go to provide zero configuration networking services to all offline devices, regardless of the platform, but we're making gradual progress.

Automatic switching to offline mode

Once all of these other areas are sorted, we'd have the technology necessary to be aware of the devices in our proximity, the services they support, and to transfer data back and forth regardless of their offline state. Advanced mesh networks exist but primarily in academia and industry, not something the average consumer can benefit from. What would a consumer-friendly mesh networking solution look like?

Obviously, we're not going to move the world away from **Domain Name System (DNS)** and DHCP anytime soon. In the same way that IPv6 has had to coexist with IPv4, we need to make sure that our applications can transition smoothly from a centralized networking model to a distributed model. In an ideal world, the underlying OS would handle this transition automatically, but in the short term, realistically, we'll need to do some of the heavy lifting in the application layer using frameworks such as Apple's Multipeer Connectivity.

The key is detecting whether we're online or offline, determined by whether we can talk to our remote Cloudant instance. If we can, we assume a centralized architecture. If we can't, we shift to a decentralized architecture. Thus, the architecture diagrams that we showed earlier are both true depending on how individual devices see themselves. It becomes interesting when you have an offline device that wants to synchronize with an online device. There are, then, three scenarios that we should consider:

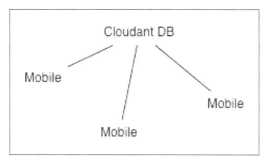

Online architecture

As this is the design that we've implemented, I won't say much about it, except that most of the time, for most users, this is the architecture that will be in use. When multiple mobile users are in the same general proximity and connected to the Internet via Wi-Fi and the network goes down, they'll be thrust into the following scenario:

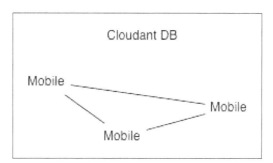

Offline architecture

As mobile devices can connect to the Internet via Wi-Fi or cellular, they will automatically switch networks and try to regain connectivity. In the meantime, any apps on these phones should try to form mesh networks to keep apps talking to each other. As phones slowly begin to connect to the Internet again, the architecture will shift yet again:

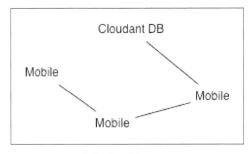

Mixed architecture

This is the most interesting diagram of the three and most representative of the real world in general. Some devices have Internet connectivity, others don't, and data can be transferred between the haves and have nots.

If the total number of devices is large and the number connected to the Internet is small, there may not be enough bandwidth to go around. In this case, the apps may decide to continue synchronizing with each other. Otherwise, the offline devices can use the online devices as a bridge to talk directly with Cloudant.

It's worth noting that this scenario is very advanced and no modern operating system handles it gracefully. This is a very common scenario in highly dense gatherings of people with sparse bandwidth but also in any scenario where a noncellular device is within the Wi-Fi range of a network with Internet connectivity.

As designers and developers, the best that we can do today is use whatever frameworks have been made available and encourage those in the industry to reach across the aisle and provide us with the tools that we need to build a truly platform-independent peer-to-peer network.

Summary

In this chapter, we explored the various ways that offline synchronization is done today. We implemented a crude version using peer-to-peer Wi-Fi and looked in depth at ways in which the industry could improve the current situation in order to allow designers and developers to deliver more compelling offline solutions.

In the next chapter, we'll end by doing a holistic evaluation of our to-do app, investigating ways to test the app in the real world and providing ideas for future improvement.

9
Testing and Measuring the UX

Congratulations! You've designed and written your first offline-first app. We've taken the principles of offline-first web development, seen how they apply to the realities of software development, and pushed the boundaries of today's technology. Keep in mind that the result of our efforts is a starting point and not the embodiment of everything an offline-first app can be. Where you go from here is up to you.

So, what's next? You've probed and prodded at the concepts, but now it's time to take what you've learned and apply it to your own applications where these lessons really matter. One reason why we explored the problems and design first is because technology is constantly changing, but human needs and behaviors remain relatively constant. By ordering the discussion in this way, you'll be able to separate the essence of the solution from the particulars of your software stack.

This said, I hope the technical details have been every bit as interesting as the softer concepts. The CouchDB ecosystem is full of fascinating, groundbreaking tools, which make offline-first development easier than ever before. Sencha Touch and Cordova are a very competent pair, making cross-platform mobile development a breeze. We stuck with the visual defaults for the most part, but Sencha Touch is quite easy to style and customize if you want. Just remember that the tools you use, while important, are never as important as the problem you're solving.

The design and implementation aside, there's one aspect of software development that we've completely ignored. Writing software is important, but setting expectations and verifying that it meets these expectations is just as important. The reason we've ignored it is because we wanted to keep each chapter as lean and focused as possible. In this chapter, we'll touch on what it means to test and measure the experience of an offline-first app.

Unlike traditional apps, we have two baselines: an offline baseline, when the Internet is not available, and an online baseline, when the Internet is available. We must test both of these scenarios and verify that the app is just as robust when online as it is when offline (or at least, anticipates the differences and handles them appropriately). We'll discuss ways to test the app and measure the improvements that we've made in a quantifiable way.

Manual testing

As you write an app, you want to verify that it works as expected. When you make changes to the app, you want confidence that you haven't inadvertently broken something.

While you can use a testing framework to verify the functionality, an easier approach, particularly for simple apps, is to exercise the UI yourself. To ensure that this is done in a consistent and reproducible way, write the test scripts for yourself or another person to follow. Let's write a few scripts now.

Creating a to-do item

The first test is to create a to-do item, add the text, maps, and an image, then verify that each of these data points have been saved. You will need two devices: one to create the to-do item and the other to verify that the data has synchronized correctly.

1. Start the app on the first mobile device and click on **Sign in**.
2. Enter your username and password and click on **Sign in**.
3. Click on **Add**.
4. Enter Test in the **Name** field and click on **Create**.
5. Click on **Edit** for the to-do list that you just created.
6. Click on **Add**.
7. Enter Test Item in the **Description** field.
8. Click on **Select** and choose an image.
9. Click on **Set**.
10. Select a point on the map.
11. Click on **Set**.
12. Click on **Create**.

At this point, you should have created a to-do list named **Test** and to-do item named **To-do Item**. The to-do item should have an image and a map location selected. Now, verify that this data has synchronized successfully:

1. Start the app on the second mobile device and click on **Sign in**.
2. Enter your username and password and click on **Sign in**.
3. Click on **Edit** for the to-do list named **Test**.
4. Click on **Edit** for the to-do item named **Item**.
5. Verify that all the data fields are present with the correct values.

Adding and removing collaborators

The second test is to take the to-do list that you created in the first test and manage collaborators. We'll add a collaborator, make a change as that collaborator, remove the collaborator, and verify that the collaborator no longer has permission to view or edit the list. You can perform this test with a single device, but multiple devices may be simpler as you can log in as a different user on each device.

1. Start the app on the first mobile device and click on **Sign in**.
2. Enter the username and password of the first user.
3. Click on **Share** for the Test to-do list.
4. Click on **Add**.
5. Enter the username of the second user in the **Name** field.
6. Click on **Share**.
7. Click on **Back**.

You've granted collaborator privileges to the second user. Now, log in as this user and verify that you can see and make changes to the list:

1. Start the app on the second mobile device and click on **Sign in**.
2. Enter the username and password of the second user.
3. Click on **Edit** for the Test to-do list.
4. Click on **Edit** for the Item to-do item.
5. Change the **Description** to **Item 2** and click on **Save**.

Now that you've made a change as a collaborator, switch back to the first device and verify that the change was synchronized:

1. On the first mobile device, click on **Edit** for the Test to-do list.
2. Verify that the name of the to-do item is **Item 2**.
3. Click on **Back**.

The next step is to remove the collaborator from the Test to-do list:

1. On the first mobile device, click on **Share** for the Test to-do list.
2. Click on **Unshare** for the collaborator that you created.
3. On the second mobile device, verify that the Test to-do list is no longer visible.

The bandwidth notifications

The final test is to make sure that the app sends appropriate notifications about the online/offline state and the quality of bandwidth during synchronization operations. You only need a single device for this test case. You may need to prime the last test step by going out of range, then retracing your steps, and repeating them:

1. Start the app on your mobile device.
2. Verify that the message **online :-)** is displayed in the footer.
3. Put the device in the airplane mode.
4. Verify that the message **offline :-(** is displayed in the footer.
5. Turn off the airplane mode and go somewhere with poor cellular reception (3G or less).
6. Create a new to-do item and add a large image.
7. Click **Create**.
8. Wait for the following messages to appear in the footer after the specified delay:

    Syncing...(immediately)

    Taking a bit longer...(10 seconds)

    Still working...(30 seconds)

    Find faster Internet? (1 minute)

    Are you on GPRS? (10 minutes?)

9. After 30 minutes have elapsed, verify that a message box appears.
10. Click **OK** to dismiss the message box.

11. Start walking away quickly from the signal.

12. When you're 10 seconds away from out of range, verify that **Going offline soon :-/** appears in the footer.

Cleaning up

To get ready for the next round of testing, open the app and delete all the lists. You'll start the next round of testing with a clean slate. These tests are not exhaustive but enough to give you a rough idea of how well the app is working. In theory, 100% test coverage is attractive but adhering to the 80-20 principle and just checking the major scenarios is probably enough.

Testing frameworks

At some point, you may decide that you're tired of running through test scripts whenever you make a code change. Fortunately, there are testing frameworks that can automate this tedious task, though writing and maintaining tests for these frameworks is its own kind of tediousness. You can use the scripts that you've already written and adapt them for whatever framework you choose. One popular option is **Bryntum Siesta**.

Bryntum Siesta

Siesta is a functional testing framework that runs in the browser. To use it, you write a series of test scripts that exercise the functionality of your app, load the framework in the browser of your choice, and run the scripts, watching them onscreen as they execute. Siesta closely emulates the experience of an actual person. It can click and input text in elements as needed.

Installing Siesta

To get Siesta up and running, we need to download Siesta Lite from Bryntum, create a harness file, and move the appropriate resources over to our app. This establishes the basic framework that we need to start writing tests:

1. Open `www.bryntum.com` in your browser.

2. Click on **PRODUCTS**.

3. Click on **Siesta**.

4. Click on **Download Siesta Lite**.

5. Under `todo-app/`, create a directory named `tests/`.

6. In this directory, create a file named `index.js`.

7. Put the following code in this file:

```
var Harness = Siesta.Harness.Browser.SenchaTouch;

Harness.configure({
  itle : 'Todo App',
  viewportWidth: 320,
  viewportHeight: 600
});

Harness.start(
  {
    group : 'Todo app tests',
    hostPageUrl : '/',
    performSetup : false,
    items : [
      // TODO: Tests go here
    ]
  }
);
```

8. Now, create another file named `index.html`, which references `index.js`.

9. Add the following code to this file:

```
<!DOCTYPE html>
<html>
<head>
<link rel="stylesheet" type="text/css" href="//cdn.sencha.com/
ext/gpl/5.1.0/packages/ext-theme-crisp/build/resources/ext-theme-
crisp-all.css" />
<link rel="stylesheet" type="text/css" href="siesta/resources/css/
siesta-all.css">

<script type="text/javascript" src="//cdn.sencha.com/ext/
gpl/5.1.0/build/ext-all.js"></script>
<script type="text/javascript" src="siesta/siesta-all.js"></
script>
<script type="text/javascript" src="index.js"></script>
</head>

<body>
</body>
</html>
```

10. Create another directory named `siesta/` under the `tests` directory.

11. Move the following files from the Siesta archive that you downloaded to the same paths under the `siesta` directory:

```
siesta-all.js
resources/
```

You're done!

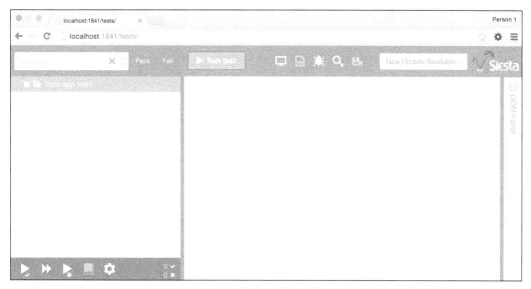

Siesta UI

Before we go any further, start your app. Open `http://localhost:8081/tests/index.html` in your browser. If the Siesta UI comes up, you're good. Commit your changes.

Writing a test

Now that Siesta is up and running, let's write a test to show how the Siesta tests work. We won't write a test for every scenario that we identified in the previous section, just the first.

Create a new file named `todo-app/tests/01_add_todo_item.t.js` and add the following code:

```
StartTest(function(t) {
  t.chain(
    t.diag("Add a todo item"),
    {waitFor: 'CQ', args: '>>todo-lists'},
```

```
{click: '>>todo-lists button[action=new]'},
{waitFor: 'CQ', args: '>>todo-list-new'},
function(next) {
   var field = t.cq1('>>textfield[name=name]');
   field.setValue('Test');
   next();
},
{waitFor: 'CQ', args: '>>textfield[value=Test]'},
{click: '>>todo-list-new button[action=create]'},
{waitFor: 'CQ', args: '>>todo-lists'},
{waitForMs: 1000},
{click: '>>todo-list-dataitem button[action=edit]'},
{waitFor: 'CQ', args: '>>todo-list'},
{click: '>>todo-list button[action=new]'},
{waitFor: 'CQ', args: '>>todo-new'},
function(next) {
   var field = t.cq1('>>textfield[name=description]');
   field.setValue('Item');
   next();
},
{waitFor: 'CQ', args: '>>textfield[value=Item]'},
{click: '>>todo-new button[action=create]'},
{waitFor: 'CQVisible', args: '>>todo-list'},
function(next) {
   t.ok(t.cq1('>>todo-list label[html=Item]'));
   next();
}
   );
});
```

Let's walk through this test step by step and I'll explain what's happening.

```
StartTest(function(t) {
```

Every test lives in a block like this, letting Siesta know where the test starts and ends. Next, we'll provide a call to `t.chain()`, which provides Siesta with the steps of the test in order.

```
t.chain()
```

This function takes a variable number of arguments, then executes them in the given order. Each argument represents a step in the test. Next, let's give the test a title.

```
t.diag("Add a todo item"),
```

This lets the user know what the test is about at a high level. Siesta supports a
Behavior Driven Development (BDD) style as well, which gives people even more
insight into what we're testing. See the Siesta documentation if you're interested in
learning more.

Next, let's provide the first actual step in the test.

```
{waitFor: 'CQ', args: '>>todo-lists'},
```

Before any test starts, we want to ensure that the application is in a good place to
start receiving commands. This statement waits for the list of to-do lists to appear.

Now that the application is ready for events, let's click something.

```
{click: '>>todo-lists button[action=new]'},
```

This statement uses a query to find an element and then clicks this element. In this
case, we're looking for the button used to create new to-do lists. Let's wait for the
next screen to show up.

```
{waitFor: 'CQ', args: '>>todo-list-new'},
```

You'll notice this pattern as we go through the rest of the test. Every time we go to a
new screen, we have to wait for this screen to appear before taking our next action.
Now that the to-do list creation screen is visible, let's fill out the form.

```
function(next) {
  var field = t.cq1('>>textfield[name=name]');
  field.setValue('Test');
  next();
},
```

This method is a single step in the test. When you have a step that's more complex
than a single statement, you can wrap it in a method. Just be sure to call `next()` at
the end of the method. All we do here is query for `textfield` called `name` and assign
it a value.

Let's confirm that the field has been set properly before creating the to-do list.

```
{waitFor: 'CQ', args: '>>textfield[value=Test]'},
```

This done, we can click on `create`.

```
{click: '>>todo-list-new button[action=create]'},
```

We wait for the lists of to-do lists to appear.

```
{waitFor: 'CQ', args: '>>todo-lists'},
```

Now, let's pause for a second before proceeding.

```
{waitForMs: 1000},
```

Normally, using `waitForMs` in a test is a bad idea because it inserts a fixed delay instead of letting the app tell us when it's ready to proceed. However, it is very useful when you're debugging a test as it allows you to slow down the test for observation purposes. I use it here just to demonstrate that use case.

Next, let's edit the to-do list that we just created and wait for the to-do list screen to become visible.

```
{click: '>>todo-list-dataitem button[action=edit]'},
{waitFor: 'CQ', args: '>>todo-list'},
```

Now, let's repeat the creation process but for a to-do item this time.

```
{click: '>>todo-list button[action=new]'},
{waitFor: 'CQ', args: '>>todo-new'},
function(next) {
  var field = t.cq1('>>textfield[name=description]');
  field.setValue('Item');
  next();
},
{waitFor: 'CQ', args: '>>textfield[value=Item]'},
{click: '>>todo-new button[action=create]'},
{waitFor: 'CQVisible', args: '>>todo-list'},
```

Once the to-do list is visible again, let's end the test by ensuring that the to-do item was created.

```
function(next) {
  t.ok(t.cq1('>>todo-list label[html=Item]'));
  next();
}
```

This method uses `t.ok()` to verify that an entry has been created for the new to-do item. If the to-do item exists, the test passes. If not, it fails.

Finally, add a reference to this test in `index.js`. Replace the to-do comment with the name of this test:

```
items : [
  '01_add_todo_item.t.js'
]
```

Now, open `http://localhost:8081/tests/index.html` in your browser. Select the test that you just created and click **Run test**. Expand the DOM panel and watch as the test executes. If the test passes successfully, you're done! If not, you'll see a screen similar to the following with a log output to help you debug:

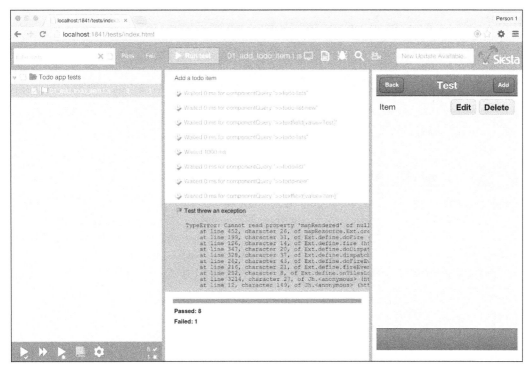

Failed test run

Commit your changes.

Before and after

Thanks to our testing, we have verified that the app is functional. However, this is not the only way to evaluate the experience. We set out to write an app that maintains the rigor of an offline-only experience while augmenting it with online-only functionality. Let's compare the before and after and see how far we've come.

Before – offline only

Open your `todo-app` folder and reset the app to the state it was in at the close of *Chapter 2, Building a To-do App*. Now, start the app and walk through the experience to refresh your memory. At this point, with a fully offline experience, you can do the following:

- Create a to-do list
- Add/remove items from the to-do lists
- Add/remove a description from a to-do item
- Add/remove an image from a to-do item
- Delete a to-do list

After – offline plus online

Now, reset the app to the state it was in at the close of *Chapter 7, Choosing Intelligent Defaults*. Walk through the same experience and note the differences. Verify that the app retains all of the functionality noted in the previous section. This functionality represents the base functionality that should operate regardless of an app's online/offline status. Test the app in both the states, offline and online.

When online, the app should additionally support the following features:

1. Add/remove a map from a to-do item.
2. Sign in with your credentials.
3. Sign out.
4. Share a to-do list with a collaborator.
5. Revoke a collaborator's privileges.
6. Synchronize to-do lists with other devices.
7. Indication of the online/offline status of the app.
8. Indication of difficulty in synchronizing the app.

There are a couple of takeaways here. First, we've successfully translated the robustness of an offline app to its online experience. This is a big deal. Not many apps can claim this.

Second, adding an online functionality is very complicated. Our initial offline attempt consisted of roughly 750 lines of MVC code. After adding the online functionality, we're now up to 2,500 lines of code. That's 333% more code, excluding tests. Code is an expensive liability. Don't underestimate the amount of time you'll need in order to get your app working properly in an online context.

This has a couple of implications. First, think carefully about whether you really need to build an online experience for the people who use your software. In the case of our to-do app, the synchronization of to-do lists is a must-have, so it was a good decision. This said, don't blindly assume that you need online capability in your application. Do your research. The second implication is that if you decide to make your app online-capable, keep the following in mind:

- Design and build the offline experience first
- Adhere to the offline-first design principles
- Test, test, test

If you keep these in mind, you'll do okay. Never underestimate the difficulty; break the problem into small discrete chunks and give yourself plenty of time to build the right thing the right way. You'll end up with an excellent experience for the people who depend on the thing that you've built.

Future improvements

We've come a long way, but we could take the experience even further. Let's talk about the current limitations of the app and some of the more complicated offline improvements that we've deferred until now.

Conflict resolution

Right now, the app can only resolve conflicts for the description of each to-do item. To do the same for maps and images, you'll need to implement a custom conflict resolution UI for each of these data types. This boils down to **diff** algorithms. You need a clear and efficient way to show the difference between an arbitrary number of different items. For text, this is relatively easy. For images and other binary data types, this is harder but not impossible. Search the Internet for image comparison algorithms if you're interested in learning more.

The cache size limits

At the moment, we cache everything. This is a trivial approach that works for our to-do app, but when you start dealing with large datasets, you'll need to think about ways to page the data in and out as needed. With PouchDB, the best way to do this is to think of databases as pages. No individual database should be able to grow to a size that your device cannot store.

Right now, we store all the text in a database, all the images in a database, all the maps in a database, and so on. In a world where storage isn't infinite, you'll need to alter this scheme. One approach would be to give each to-do list or to-do item its own database. In CouchDB, databases are cheap, so this is actually an effective strategy. In more traditional SQL databases, you might choose to map each to-do list to a table instead.

Account management

We avoided a lot of code by relying on Cloudant's built-in user management. Unfortunately, this means that user accounts must be set up manually. Any serious app must include account management capabilities. This can either be done in-app or on an external website. The former is a better experience but the latter may be required anyway. For example, if you set up an account and then lose your phone, you need a way to reset your password or view your data without access to the app.

Offline Maps

Another limitation is that our use of Google Maps requires an Internet connection. If you start the app while offline, Google Maps will refuse to load. This is very poor behavior. HTML5 introduces the concept of an application cache. If you specify a cache manifest, the browser will use the local copy of any specified files and update automatically when the remote copy changes.

To implement this, you need to create a cache manifest and add every URL used by Google Maps. If you plan to use the app with Firefox or Chrome — browsers notorious for their aggressive caching behavior — you may need to override how they cache your app as well.

> Alternatively, you can use the **Google Static Maps API** to cache maps as images, which can circumvent this problem but at a higher storage cost. For more information about static maps, see the following link:
>
> ```
> https://developers.google.com/maps/documentation/
> static-maps/
> ```

Granular sync indicators

Right now, the app tells you when the synchronization is happening. This is fine, but it's also useful to tell people which lists or items are being synchronized. If I see that my grocery list is being updated, I might hold off making changes to it until I see how it changed. Similarly, when I first start the app, I want the app to load its own configuration and metadata before showing me a potentially outdated UI. We haven't implemented a splash screen, but ordinarily you'd execute these initial loading operations while the splash screen is displayed.

Offline sync

In the last chapter, we discussed the current state of cross-platform peer-to-peer synchronization. It's not great. Your best bet is to use peer-to-peer Wi-Fi, but even that path is fraught with peril. To improve in this area, the major mobile device manufacturers will need to get behind industry standards that allow mobile devices to easily share data and form mesh networks, which allow groups of devices to retain elements of the online experience.

Summary

In this chapter, we wrote tests to verify the functionality of the to-do app and showed how you can turn these manual tests into automated tests to be run by an appropriate testing framework. We also did a holistic evaluation of the before and after state of our app, showing how far we've come while preserving the rigor of the offline experience. Finally, we briefly described ways in which the experience could be improved further.

Take a minute and pat yourself on the back. Now, think about how to apply the lessons that you've learned to your own software projects. I'd love to hear how you apply what you've learned. Feel free to drop me a line at offlinefirstux@gmail.com or tweet me at @offlinefirstux. Best of luck!

References

We've supplied some links for further reading about the topics in each chapter. You may find these useful to learn more about offline-first web development.

Chapter 1

- **ICT Facts & Figures**: The World in 2015: `http://www.itu.int/en/ITU-D/Statistics/Documents/facts/ICTFactsFigures2015.pdf`

- **Users Have Low Tolerance For Buggy Apps**: `http://techcrunch.com/2013/03/12/users-have-low-tolerance-for-buggy-apps-only-16-will-try-a-failing-app-more-than-twice/`

Chapter 2

- **Sencha Touch**: `https://www.sencha.com/products/touch/`

- **Sencha Touch 2.4.2 API Docs**: `http://docs.sencha.com/touch/2.4/2.4.2-apidocs/`

Chapter 3

- **Designing Offline-First Web Apps**: `http://alistapart.com/article/offline-first`

- **Offline Patterns**: `https://www.ibm.com/developerworks/community/blogs/worklight/entry/offline_patterns?lang=en`

- **How Can We Make Mobile Apps Suck Less Offline**: `https://gigaom.com/2010/10/15/how-to-make-mobile-apps-suck-less-offline/`

- **Dealing with Disconnected Operation in a Mobile Business Application: Issues and Techniques for Supporting Offline Usage**: `http://bricklin.com/offline.htm`
- **Google Maps API**: `https://developers.google.com/maps/`

Chapter 4

- **PouchDB**: `http://pouchdb.com`
- **remoteStorage**: `https://remotestorage.io`
- **Hoodie**: `http://hood.ie`
- **IBM Cloudant**: `https://cloudant.com`
- **Offline First community**: `http://offlinefirst.org`

Chapter 6

- **A plain English introduction to CAP Theorem**: `http://ksat.me/a-plain-english-introduction-to-cap-theorem/`
- **Errors in Database Systems, Eventual Consistency, and the CAP Theorem**: `http://cacm.acm.org/blogs/blog-cacm/83396-errors-in-database-systems-eventual-consistency-and-the-cap-theorem/fulltext`

Chapter 7

- **Caching in the Distributed Environment**: `https://msdn.microsoft.com/en-us/library/dd129907.aspx`

Chapter 8

- **Apple MFi Certification**: `https://developer.apple.com/programs/mfi/`
- **Bug preventing iOS and Android devices from pairing**: `https://code.google.com/p/android/issues/detail?id=58725`
- **PhoneGap plugin for NFC on Android**: `https://github.com/chariotsolutions/phonegap-nfc`
- **PhoneGap plugin for Wi-Fi on Android**: `https://github.com/HondaDai/PhoneGap-WifiInfoPlugin`

- **Apple Multipeer Connectivity Framework**: `https://developer.apple.com/library/ios/documentation/MultipeerConnectivity/Reference/MultipeerConnectivityFramework/`

- **Android Connectivity API**: `http://developer.android.com/guide/topics/connectivity/`

- **CouchDB**: `http://couchdb.apache.org`

Chapter 9

- **Google Static Maps API**: `https://developers.google.com/maps/documentation/static-maps/`

- **Bryntum Siesta**: `http://www.bryntum.com/products/siesta/`

- **Bryntum Siesta API docs**: `http://www.bryntum.com/docs/siesta/`

Index

A

active communication 159
Airdrop 235
Apple Safari
  cache, clearing 230

B

Behavior Driven Development (BDD) 273
Bluetooth 248-251
Bonjour 253
browser
  using 16, 17
Bryntum Siesta
  about 269
  installing 269-271
  test, writing 271-275
  URL 269

C

cache
  clearing 229
  clearing, in Apple Safari 230
  clearing, in Google Chrome 230
  clearing, in Microsoft Edge 230
  clearing, in Mozilla Firefox 230
  limitations 228, 229
Chrome
  installing 16, 17
  URL 16
Cloudant DB 247
command line 18
communication
  active communication 159
  passive communication 159

compensation, for bad network conditions
  databases, creating 172
  follow-up communication, sending 190
  Item store, updating 175
  list store, updating 183
  Main controller, refactoring 172-174
  models, updating 184
  new users, creating 172
  performing 169-171
  Sync controller, wiring up 185
  views, updating 184
conflict detection
  conflicts, attaching to item model 207
  conflicts, returning with PouchDB 206
  current revision, attaching to
     item model 208
  implementing 206
  revision, checking 208, 209
conflict resolution
  controller logic, adding 213-215
  fields, adding to edit views 211-213
  for images 216
  for maps 216
  implementing 210
  supporting methods, adding 210
  to-do item, defining 215, 216
Cordova camera plugin
  installing 82, 83
CouchDB 73
Craigslist 11
Cross-Origin Resource Sharing (CORS) 112

D

design principles
  comparing 152, 153
  contrasting 152, 153

sync, restricting to users lists 198
updating, for bad network conditions 175

J

Java
  installing 19, 20

L

last view, offline-first application
  model, creating 224, 225
  page, loading 226-228
  pages, specifying to restore 225, 226
  restoring 223, 224
  store, creating 224, 225
least developed countries (LDCs) 10
list frequency 229
lists, sharing
  existing views, modifying 144
  list store, modifying 147
  logic, adding to controller 148-151
  model, adding 145
  share views, creating 142, 144
  store, adding 146
list store
  databases, flagging 184
  pointer, creating for metadata database 183
  updating, for bad network conditions 183
LocalStorage 110, 114

M

manual testing
  about 266
  bandwidth notifications, sending 268
  collaborators, adding 267, 268
  collaborators, removing 267, 268
  to-do item, creating 266, 267
  to-do item, deleting 269
map documents 171
mapping support
  controller, wiring up 93-95
  custom map component, creating 88, 89
  drilldown support, adding
    to controller 97-99
  implementing 87
  logic, adding to view 89-93

map view, building 95-97
map views, refactoring 102-104
map view, wiring up 99-101
MemoryProxy 111
metadata documents 171
MFi certification
  reference link 251
Microsoft Edge
  cache, clearing 230
Model View Controller (MVC) 145
Mozilla Firefox
  cache, clearing 230
Multipeer Connectivity
  about 251
  reference link 251
multiple lists, online-only features
  backing store, creating 133-136
  logic, adding to controller 139-142
  managing 128
  model, creating 139
  sync logic, removing from
    item store 137, 139
  views, implementing 129-133
  views, refactoring 129

N

Near Field Communication (NFC) 248-251
npm
  about 111
  installing 19

O

offline 244-247
offline databases
  about 110
  Hoodie 113
  PouchDB 111
  remotestorage.io library 112
  Sencha Touch 110, 111
offline design
  principles 7-10
offline experience
  app design 29
  data, safeguarding 29
  delays, avoiding 28
  designing 25

PouchDB
about 12, 111, 249
adding, to app 114
updating 115-117
URL 113
predictive algorithm
additional points, adding 161, 162
existing points, updating 162
future connectivity, predicting 163
location, ensuring 161
need for 159
online/offline state, setting 164
position store, creating 164, 165
seed location, creating 161
velocity, ensuring 163
writing 160

R

remotestorage.io library 12, 112
Representational State Transfer (REST) 255

S

Secure Shell (SSH) 249
Sencha Touch application
about 110, 111
changes, committing 41
configuring 21-23
create view, adding 37, 38
creating 33
edit view, adding 39-41
files, generating with Sencha cmd 33, 34
list view, adding 35, 36
main view, creating 34, 35
URL 21
Server 244
split-brain
about 194
collective agreement 195, 196
self-appointed dictator 197, 198
Sublime Text 3
about 15
installing 15, 16
URL 16
Sync controller
connection logic, tweaking 186
databases, initializing 185

synchronization, modifying 187-189
synchronization, prioritizing 189, 190
wiring up, for bad network conditions 185
synchronization, over Wi-Fi
about 252
IP address, assigning to device 253
P2P connection, setting up 252, 253
support, implementing for
offline syncing 255-257
system state
error messages, displaying 158
exposing, to user 155
online/offline indicator, creating 156

T

tags 250
testing frameworks
Bryntum Siesta 269
text documents 171
text editor 15, 16
to-do app
evaluating 275
with offline experience 276
with offline plus online experience 276, 277
to-do app, improvements
account management 278
cache size limits 277
conflict resolution 277
granular sync indicators 279
offline Maps 278
offline sync 279

U

up front communication
setting up 159
USB 250, 252
user account
controller references, updating 122, 123
credentials, generating 119
handler functions, adding 124-127
item store, tweaking 127, 128
managing 119
sign in view, creating 121, 122
stores, expanding 120
user guidance
modal, adding to app 168

Thank you for buying
Offline First Web Development

About Packt Publishing

Packt, pronounced 'packed', published its first book, *Mastering phpMyAdmin for Effective MySQL Management*, in April 2004, and subsequently continued to specialize in publishing highly focused books on specific technologies and solutions.

Our books and publications share the experiences of your fellow IT professionals in adapting and customizing today's systems, applications, and frameworks. Our solution-based books give you the knowledge and power to customize the software and technologies you're using to get the job done. Packt books are more specific and less general than the IT books you have seen in the past. Our unique business model allows us to bring you more focused information, giving you more of what you need to know, and less of what you don't.

Packt is a modern yet unique publishing company that focuses on producing quality, cutting-edge books for communities of developers, administrators, and newbies alike. For more information, please visit our website at www.packtpub.com.

Writing for Packt

We welcome all inquiries from people who are interested in authoring. Book proposals should be sent to author@packtpub.com. If your book idea is still at an early stage and you would like to discuss it first before writing a formal book proposal, then please contact us; one of our commissioning editors will get in touch with you.

We're not just looking for published authors; if you have strong technical skills but no writing experience, our experienced editors can help you develop a writing career, or simply get some additional reward for your expertise.

Responsive Web Design with HTML5 and CSS3

Second Edition

ISBN: 978-1-78439-893-4   Paperback: 312 pages

Build responsive and future-proof websites to meet the demands of modern web users

1. Learn and explore how to harness the latest features of HTML5 in the context of responsive web design.

2. Learn to wield the new Flexbox layout mechanism, code responsive images, and understand how to implement SVGs in a responsive project.

3. Make your pages interactive by using CSS animations, transformations, and transitions.

Bootstrap Essentials

ISBN: 978-1-78439-517-9   Paperback: 166 pages

Use the powerful features of Bootstrap to create responsive and appealing web pages

1. Learn where and how to use Bootstrap in your new web projects.

2. Design and develop mobile first web portals that support all devices.

3. A step-by-step guide with easy-to-follow practical exercises to develop device friendly websites.

Please check **www.PacktPub.com** for information on our titles

Building a Responsive Site with Zurb Foundation [Video]

ISBN: 978-1-78328-517-4 Duration: 01:53 hours

A fascinating journey from the very basics of Zurb Foundation to a complete responsive website

1. Use grid layouts effectively to create a consistent experience on all mobile devices and desktops.

2. Beautify your page and enable automatic content structuring using accordion and equalizer.

3. Program your site to adapt to any screen size using interchange and specific lightweight content.

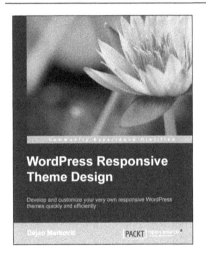

WordPress Responsive Theme Design

ISBN: 978-1-78528-845-6 Paperback: 228 pages

Develop and customize your very own responsive WordPress themes quickly and efficiently

1. Structured learning for new developers and technical consultants to enable you to build responsive WordPress themes.

2. Concise and easy-to-follow walkthroughs of WordPress, PHP, and CSS code.

3. Packed with examples and key tips on how to avoid potential pitfalls.

Please check **www.PacktPub.com** for information on our titles